Springer Remote Sensing/Photogrammetry

The Springer Remote Sensing/Photogrammetry series seeks to publish a broad portfolio of scientific books, aiming at researchers, students, and everyone interested in the broad field of geospatial science and technologies. The series includes peer-reviewed monographs, edited volumes, textbooks, and conference proceedings. It covers the entire area of Remote Sensing, including, but not limited to, land, ocean, atmospheric science and meteorology, geophysics and tectonics, hydrology and water resources management, earth resources, geography and land information, image processing and analysis, satellite imagery, global positioning systems, archaeological investigations, and geomorphological surveying.

More information about this series at http://www.springer.com/series/10182

M. B. Rajani

Patterns in Past Settlements: Geospatial Analysis of Imprints of Cultural Heritage on Landscapes

Springer

M. B. Rajani
National Institute of Advanced
Studies (NIAS)
Bengaluru, India

ISSN 2198-0721 ISSN 2198-073X (electronic)
Springer Remote Sensing/Photogrammetry
ISBN 978-981-15-7465-8 ISBN 978-981-15-7466-5 (eBook)
https://doi.org/10.1007/978-981-15-7466-5

This Springer imprint is published by the registered company Springer Nature Singapore Pte Ltd.
The registered company address is: 152 Beach Road, #21-01/04 Gateway East, Singapore 189721, Singapore

To my mentors
John R. Marr and K. Kasturirangan
for challenging me
never to stop learning

Foreword by Ranganath Navalgund

The advent of geospatial techniques in the last few decades has greatly enriched investigations in the field of landscape archaeology. The emergence of this field, sometimes referred to as "Geospatial Archaeology", is particularly of interest in the Indian context, which presents a diverse network of cultural heritage sites/networks ranging in time from the Palaeolithic to Early Historic periods to ancient, medieval, and modern periods. Geospatial techniques comprise remote sensing of landscapes from various platforms, geographic information systems, and global navigation satellite systems. Remote sensing (RS) of the earth's surface in visible, infrared, thermal, and microwave regions provides a synoptic view of the area of interest. RS data in different wavelength regions captures characteristic signatures associated with potential archaeological sites at different levels of details and helps in their identification, delineation, and further exploration. Geographic information system (GIS) facilitate the integration of spatial data from different sources with attribute data related to archaeological sites. GIS tools can also be used for spatial analysis and modelling, which can be valuable for further documentation and conservation of sites and their environs. Global navigation satellite systems provide precise coordinates of each point to assist in geotagging archaeological sites for further analysis and exploration.

While there have been a few preliminary studies in Geospatial Archaeology, the field has dramatically blossomed over the last decade and a half. Rajani and her colleagues have been at the forefront of its systematic development, and this book is a culmination of their splendid and painstaking efforts. One of the key contributions of this book, in my view, is an articulation of the logical connections between the hypothetical changes that an archaeological site might undergo as a result of various morphological processes over time and under different conditions, and how well these changes are captured by the satellite images through evolving signatures. The scientific basis for archaeological signatures has been described by judiciously balancing rigour and readability, complemented by helpful illustrations. Readers without a background in remote sensing will also find clear explanations of the relevant remote sensing principles and terminology. This book also demonstrates the power of GIS techniques to collate ancient and modern maps, hand-drawn schematics, paintings, and satellite images into a single platform to understand the possible evolution of and relationships between archaeological features. Rajani presents a synthesis of these core ideas through a number of compelling case studies, viz. the Harappan civilization, the Buddhist site of Nalanda, Agra, Talakadu, Srirangapatna, and the Seven Pagodas at Mahabalipuram. These examples clearly demonstrate the methodical approach to be adopted by any researcher attempting to use RS and GIS in unravelling the mysteries of archaeological sites in their complex and varied contexts. Lastly, this book objectively examines the enormous challenges of documenting and conserving archaeological sites. The challenges are particularly complex in the Indian context, given the sheer volume of sites and the conflicts arising from the need to protect rich heritage on the one hand and national developmental issues on the other. Rajani presents a cogent strategy to help address these challenges, backed by concrete examples.

Geospatial Archaeology is relatively a new discipline, and there is dearth of textbooks or monographs which can introduce this field to young researchers in archaeology and history, as well as working professionals including archaeological surveyors, planners, and policy-makers. I am sure that this book will be a valuable introduction for each of these readers to the nuances of technological tools for archaeological research and will help them appreciate how well these tools can be used in a very practical sense.

Bengaluru, India
May 2020

Ranganath Navalgund
Former
Vikram Sarabhai Distinguished
Professor
ISRO, Bengaluru

Director
Space Applications Centre (ISRO)
Ahmedabad and National Remote
Sensing Centre (ISRO), Hyderabad

Foreword by Frederick Asher

I am honoured to write this forward to M. B. Rajani's *Patterns in Past Settlements*. At the same time, I cannot think of anyone less qualified to write this than I am. I am an art historian and a humanist, and this is a book of science, a wonderfully rich and important one that I am so grateful to have read. My own work has benefitted substantially from Rajani's remote sensing insights. No longer can or should art historians rely solely on their close inspection of works of art, large structures, or whole sites. For some decades, X-ray analysis has benefitted the study of paintings, and carbon-14 studies have helped us assign dates to works with sufficient organic material to make such analysis possible. But more recently, science has pushed us a great deal farther, probably much farther than most humanists realize. Looking at the examples from Rajani's own research that she cites in this book, I realize the extent to which geospatial analysis can provide insights we never would have imagined just a decade or two ago. The insights give us a sense of how people in the past thought, for example, as they laid out temples oriented towards a specific star. They show us things, physical remnants of the past, that we could not see with our eyes because they lie underground or beneath the sea. And they provide tools for understanding relationships that we might not imagine. This book, maybe most importantly, blurs the distinction between science and humanities or rather shows how the two can work collaboratively, each enriching the understandings of the other. In a world that is ever more closely connected by instantaneous communication, Rajani has here shown how disciplinary boundaries can be transcended to advance human knowledge. I was once certain that I knew precisely the boundaries of my discipline, art history. I no longer know. I no longer care. That is because my field, like so many others, is ever so much richer because we are breaking barriers, as Rajani demonstrates so effectively in this volume. As Romila Thapar has argued recently, "Historical studies today recognize many new ways of analysing the past".[1] She means, in her words, "the use of other social sciences in

[1]Romila Thapar, "A Response to the Conference". In Kumkum Roy and Naina Dayal, eds., *Questioning Paradigms/Constructing Histories; A Festschrift for Romila Thapar*. New Delhi: Aleph Book Company, 2019, p. 466.

suggesting methods of analysis and the use of the past in legitimizing the present". But I think we need to add science to the mix, and it requires someone like Rajani who can transcend the sciences, social sciences, and humanities to show us the way. That is precisely what this volume does.

May 2020

Frederick Asher
Professor Emeritus of Art History
University of Minnesota
Minneapolis, USA

Preface

With or without our consent, details from our daily lives are being captured and stored in vast digital databases. If humanity were to vanish tomorrow and alien historians chanced upon these databases centuries hence, they would have incredibly rich resources for studying humanity's past. As humans in the twenty-first century trying to understand our own past, we have no comparably detailed database to tap into. Instead, we must scavenge for traces of past human activity that have survived in material form and develop expertise in interrogating these materials before their secrets fade into oblivion. To scrutinize written records, we turn to historians. We rely on archaeologists to scrutinize ancient materials ranging from small-scale fragments of pottery and brick to larger-scale artefacts such as individual buildings and whole settlements. In each of these domains, scholars have benefitted from scientific developments that are constantly expanding the kinds of ancient materials that are interrogable, and technological advances that enhance our ability to extract useful information from them.

This book is an introduction to a new branch of archaeology that scrutinizes landscapes. Humans necessarily scar the landscape when they construct structures, forts, canals and road networks. These scars, which can survive for many centuries, are also material records of past human activity and are therefore worthy of study. However, these scars are difficult to discern at ground level, and their value was largely unrecognized until they were observed from an aerial perspective, initially from aircraft and more recently from satellites. Apart from such images, analysts can pull together spatial data from other sources including old maps, paintings, textual reports, and field surveys. This coherent spatial view allows an analyst to hunt for visual patterns in a landscape in a manner akin to an art historian examining a painting through an appropriate historical lens. Furthermore, just as a modern art historian uses technology to scrutinize what lies beneath the topmost layer of paint, advances in the technology of remote sensing (RS) permit satellites to view landscapes in different wavelengths and reveal hidden layers of human artefacts.

The purpose of such analyses is to identify, understand, conserve, and protect remnants of our past. This field therefore brings together knowledge from archaeology, history, art history, architectural history, heritage science, geoarchaeology, geomorphology, remote sensing, geographic information system (GIS), and cultural heritage conservation and management. Although I am not an expert in all these domains, my research group has brought elements from each of these domains to bear in investigating landscapes surrounding several archaeological sites, and we have had the immense satisfaction of contributing to our overall understanding of these sites. While all these sites are situated in the Indian subcontinent, my research has helped me recognize and articulate universal principles that I am excited to present in this book.

I hope this book will be of interest to at least three groups of readers. The first group comprises students and budding or established researchers in fields such as archaeology, history, and art history who see potential value in analysing the landscape in their region of interest. Readers in this group typically have little or no exposure to the fundamentals of RS and GIS, and I hope that this book will flatten the seemingly daunting learning curve that they face. I was once part of this group, and I now recognize that my initial hesitancy in adopting RS and GIS was rooted in a cultivated fear of science and technology. I was fortunate to have mentors to guide me past this mental hurdle, and I hope that my book will similarly draw curious minds from this first group into this fascinating field of research by providing them an accessible pathway.

The second group of readers who I hope to interest are authorities concerned with the protection of cultural heritage. Readers from this group are part of international organizations such as UNESCO, as well as national and subnational agencies. In India, these agencies include the Archaeological Survey of India (ASI) and each State's Department of Archaeology. Their mandate includes defining property boundaries for protecting archaeological sites and regulating changes to land use in their vicinity. This book aims to stimulate discussions among this community of professionals on how best practices in their field should be updated in response to this new branch of archaeology. This book will also propose new models of collaboration between these agencies and academia that could lead to more effective urban planning without compromising the protection afforded to sites.

Finally, I hope this book will appeal to readers from non-governmental bodies and private professionals who are involved in making maps and master plans for cultural heritage protection and conservation. I hope that the techniques presented in this book will help readers from this group develop more efficient and effective processes as they scramble to document and protect thousands of immovable cultural heritage artefacts before these are consumed by expanding cities, industries, transportation networks, and other hallmarks of rapid development.

In organizing this book, I have tried to keep in mind the differing expertise and demands of these three groups of readers. Chapter 1 is an introduction to the field and should be of interest to all readers. I end this chapter with an overview of the rest of the book so that readers can skip ahead to specific chapters of interest if they wish. (To help readers, most chapters start with a Preamble that summarizes its main purpose.) For instance, readers from the latter two groups may wish to jump directly from Chaps. 1 to 6. The latter chapters have several references to specific sections within the foundational chapters (Chaps. 2–4), and readers can look up or refresh their understanding of these foundational concepts as needed. While I expect readers to pay greatest attention to applications, I am certain that some will be curious about the scientific basis underlying these applications. I have placed such material in Boxes to avoid disrupting the main flow of ideas, while remaining readily accessible.

Throughout this book, I have referenced sources of satellite imagery including ISRO's Bhuvan portal and the USGS portal. Steps to access these resources are excluded, since up-to-date tutorials are readily available from their respective websites. Similarly, it is beyond the scope of this book to provide detailed guidance on using RS and GIS tools. Instead, I will try to excite your interest in the domain to such an extent that you will be eager to pick up the necessary skills from available instructional materials. I hope you will join us in an endlessly fascinating examination of our rich heritage, aided by technology, and the vital task of preserving as much of it as we can for posterity.

Bengaluru, India M. B. Rajani

Acknowledgements

I began conceiving such a book as I was finishing my Ph.D. (2010–2011), but the project did not gain impetus until I was selected for a Homi Bhabha Fellowship that supported me from November 2014 to January 2017. This Fellowship further supported an academic visit to the UK as well as several domestic trips to interact with students, and for field and archival work, all of which have proved invaluable. Although it has taken more than three additional years to complete this project, I can say with confidence that this book is substantially richer for the delay, and I remain grateful for this support. I am deeply thankful to late Prof. B. V. Sreekantan, who encouraged me to apply for this Fellowship. He was always keenly interested in my progress and would have been glad to see this book completed. I am grateful to Prof. S. M. Chitre, Hon. Executive Director of the Homi Bhabha Fellowships Council, for his continued encouragement.

The National Institute of Advanced Studies (NIAS), Bengaluru, has provided me a strong platform for conducting multidisciplinary research, as a Ph.D. student (2005–2010), as a postdoc (2011–2012), and now as a faculty member (from 2015). I am thankful that my institution has provided me so many opportunities for personal growth and to build a group of researchers, students, and colleagues. I thank Directors: Dr. K. Kasturirangan (who was also my Ph.D. supervisor), Dr. V. S. Ramamurthy, late Prof. Baldev Raj, and Dr. Shailesh Nayak for not only creating an enriching ambience at NIAS, but also for their mentorship and support. NIAS's milieu has provided unique opportunities for collaborations with Prof. Roddam Narasimha, Prof. S. Ranganathan, Prof. Sharada Srinivasan, Prof. Dilip Ahuja, Prof. Anindya Sinha, Prof. Sindhu Radhakrishna, Prof. Srikumar Menon, Prof. Nithin Nagaraj, Dr. Shalini Dixit, Dr. V. V. Binoy, Prof. Ravi Korisettar, and Prof. H. S. Mukunda. I also cherish interactions with senior colleagues who have sadly passed away, including Prof. B. V. Sreekantan, Prof. S. Settar, Prof. P. G. Vaidya, and Prof. Tim Poston. I am grateful to each of them for their inputs and to colleagues and collaborators outside NIAS whose names appear as co-authors in papers referenced throughout this book. I warmly thank Dr. Brinda Kumar, Assistant Curator at The Metropolitan Museum of Art, New York, who provided access to several books, maps, and other invaluable resources that I would have otherwise struggled to lay hands on.

My research has been supported by the Government of India through the Indian Space Research Organisation (I would specifically like to acknowledge the interest that Dr. P. G. Diwakar and Dr. Prakash Chauhan have taken in an ongoing project to initiate a National-level Geospatial Database for built heritage, the implications of which are detailed in Chap. 6), Science and Engineering Research Board, Ministry of Earth Sciences (I would like to acknowledge the interest and encouragement provided by Prof. Satish Shetye and Cdr. P. K. Srivastava), the Department of Science and Technology (I am grateful for inspiring in-person and online interactions with Dr. Ashutosh Sharma and to Dr. Santanu Chaudhury who has led DST's initiative on Indian Heritage in Digital Space), and the Government of Karnataka through the Karnataka Knowledge Commission (KKC) (I acknowledge the interest and encouragement provided by Dr. Mukund Rao, Prof. S. Sadagopan, and the then Commissioners of the Department of Archaeology, Museums and Heritage (DAMH), Karnataka, Dr. C. G. Betsurmath and Dr. N. Manjula). I thank these organizations for the generous research grants that have allowed me to build a group of researchers and develop a rich repertoire of case studies.

I am grateful for the support I have received from specific agencies while working on certain sites and projects: the DAMH, Karnataka, with whom I first collaborated at Talakadu (I would like to specifically thank Prof. M. S. Krishna Murthy who led the Talakadu excavations; Dr. A. S. Rajawat from Space Applications Centre, Ahmedabad, for his collaboration; and Dr. R. Gopal and Dr. T. S. Gangadhar, the then Director and Deputy Director of DAMH), Nalanda University, which awarded me a 9-month Fellowship that enabled me to begin work at Nalanda (I am grateful to the then Vice Chancellor Dr. Gopa Sabarwal and Dean Dr. Anjana Sharma), the Bihar Heritage Development Society with whom I collaborated at Nalanda and other sites in Bihar (I thank Dr. Bijoy Chaudhury for his continued support and interest), and IIT Gandhinagar which supported my work at Dholavira, where I had several fruitful interactions with Prof. Michel Danino and Dr. V. N. Prabhakar from the Archaeological Survey of India (ASI). I am also grateful to ASI for consultations, interactions, and assistance during site visits and for numerous opportunities to showcase my work.

I thank the anonymous reviewers of Springer Remote Sensing/Photogrammetry series whose feedback about my book structure was immensely helpful. The conceptualization of such a book and the pedagogic sequencing of ideas were greatly influenced by the opportunities I have had to interact with students in allied fields. I thank Sri Rajendra Pawar for inviting me to set up and teach the M.Tech. GIS programme at NIIT University, Neemrana. Similarly, I thank Karnataka State Remote Sensing Applications Centre (KSRSAC) for inviting me to teach in their M.Tech. programme for two semesters. I have conducted short courses and hands-on workshops for master-level students with the kind support of Prof. Michel Danino (IIT Gandhinagar) and Prof. S. Sadagopan (IIIT Bangalore). I thank Dr. B. S. Shylaja and Dr. Madhusudhan for providing me opportunities to present my work at the annual Research Education Advancement Programme (REAP) of

Jawaharlal Nehru Planetarium, Bangalore. Similarly, I am grateful to Prof. Upinder Singh and Prof. Nayanjot Lahiri who invited me to interact with students from the fields of humanities and social sciences at the History Department of their former institution, Delhi University. I thank Thatha Gatha Neogi (formerly at Kerala Council for Historical Research) and Dr. Sanjukta Dutta (Ashoka University) for providing me similar opportunities, Krupa Rajangam for arranging such an interaction at Dayananda School of Architecture, and Dr. Deborah Swallow for arranging interactions with students at the Courtauld Institute of Art. I owe the success of these interactions in large part to detailed and critical feedback on initial and refined versions of my presentations from the researchers and students who have been working with me: Kuili Suganya, Ekta Gupta, Sonia Das, and Dr. Asmita Mohanty.

I am deeply indebted to my mentors who have maintained faith in my ability to explore a new domain and have been extremely generous with their time to guide me throughout my career. Here, I specifically acknowledge those who helped me in compiling this book: I thank Dr. George Joseph, whose advice in organizing the contents has been invaluable. I am deeply grateful to Dr. Ranganath Navalgund, who has known me since my first foray into this domain and who has exhorted me to write this book as soon as I completed my Ph.D. He has very kindly reviewed the entire manuscript and has written a foreword as well. It has been a joy to collaborate with Prof. Frederick Asher on the sites of Nalanda and Sarnath. His excellent advice has helped me showcase my work among the wider international community of art historians, archaeologists, and scholars from allied fields. I also thank him for agreeing to review the full manuscript and for endorsing it with his generous foreword. The funds that he and Prof. Catherine Asher provided have supported my work on Sarnath and Mehrauli.

I warmly thank my colleagues Ekta Gupta and Kuili Suganya. They have not only been research collaborators in some of the case studies presented here, but painstakingly reviewed the technical content and helped finalize the figures. Finally, I thank my husband Viraj Kumar, not only for his constant support, but for serving as a sounding board to discuss high-level ideas and effective ways to present them. Since his expertise lies outside the realms of remote sensing, GIS, and heritage studies, he has critically reviewed the manuscript from a layperson's perspective, and his inputs have greatly improved its readability.

Contents

Acronyms and Abbreviations

ASI	Archaeological Survey of India
B/H	Base Distance/Height
DEM	Digital Elevation Model
DN	Digital Number
DTM	Digital Terrain Model
GCP	Ground Control Point
GEOBIA	Geographic Object-Based Image Analysis
GIS	Geographic Information System
GNSS	Global Navigation Satellite System
GPS	Global Positioning System
INTACH	Indian National Trust for Art and Cultural Heritage
LiDAR	Light Detection and Ranging
NDVI	Normalized Difference Vegetation Index
NIR	Near Infrared
RS	Remote Sensing
SAR	Synthetic Aperture Radar
Sat nav	Satellite Navigation
SOI	Survey of India
UAV	Unmanned Aerial Vehicle
UNESCO	United Nations Educational, Scientific and Cultural Organization
WHC	World Heritage Convention

Chapter 1
Introduction

Preamble

UNESCO's *Convention Concerning the Protection of the World Cultural and Natural Heritage* provides a broad definition of cultural heritage that includes monuments, groups of buildings and archaeological sites of outstanding universal value (Article 1).[1] This definition of cultural heritage could arguably be extended to include smaller-scale *tangible expressions* such as tools, weapons, sculptures, murals as well as *intangible expressions*—traditional skills, crafts, folklore, rituals, etc. Our primary focus will be on cultural heritage as per UNESCO's definition, which we will call *built heritage* to avoid confusion with broader interpretations.

The Convention recognizes that built heritage is "increasingly threatened with destruction not only by the traditional causes of decay, but also by changing social and economic conditions which aggravate the situation with even more formidable phenomena of damage or destruction". Hence, when structures at a site are identified as built heritage, the Convention seeks to establish a system for their protection. These structures constitute the *known* remnants of built heritage at the site, and it is possible that hitherto *unknown* remnants associated with the site still exist in its vicinity. The central thesis of this book is that a careful study of a site's vicinity or *spatial context* can lead to two important types of discoveries that may improve our overall understanding of the site and inform our strategies for its protection: the discovery of further instances or attributes of built heritage, and the discovery of artefacts such as former water bodies, canals, and refuse mounds associated with past human activity at the site.

Today, such studies are conducted with the help of a rich array of technical tools. Our intention is to help readers understand how to leverage these technologies to conduct such studies. In doing so, we will detail several studies that have not only

[1] https://whc.unesco.org/en/conventiontext/. Accessed on 12 Apr 2020.

M. B. Rajani, *Patterns in Past Settlements: Geospatial Analysis of Imprints of Cultural Heritage on Landscapes*, Springer Remote Sensing/Photogrammetry,
https://doi.org/10.1007/978-981-15-7466-5_1

Fig. 1.1 Locations of primary sites discussed in this book

yielded new site-specific scholarly insights but have assisted efforts to preserve built heritage. Many of the examples presented in this book are drawn from our own research, which has primarily focused on sites in the Indian subcontinent. Figure 1.1 shows the locations of these sites. Nevertheless, the techniques we present are widely applicable.

In this introductory chapter, we first briefly review the history of studying the spatial contexts of sites and place it within the wider discipline of archaeology. Next, we review the development of tools that help archaeologists interrogate spatial contexts ever more effectively to yield clues about their past. Finally, we discuss the relative advantages of space-based platforms over aerial platforms in the search for archaeological clues.

1.1 The Study of Spatial Contexts: A Brief History

Every culture has developed myths and legends that reflect a deep human urge to investigate our own origins, and the origins of our societies and practices. Ancient monuments and antiquities have naturally been prime objects of interest in such investigations, and the Greek historian Herodotus (fourth century BCE) was the first to bring a logical and sceptical approach to the study of artefacts already recognized as ancient in his time (Herodotus 2006). The broad discipline of archaeology (literally, "the study of ancient things") is concerned with systematically studying physical material evidence of past societies to answer questions such as: How did people organize themselves into social groups and exploit their surroundings? What did they eat, make, and believe? How did they communicate? Why did their societies change? (Renfrew and Bahn 1996).

Like many disciplines, archaeology matured rapidly during the Renaissance movement in Europe (fourteenth to seventeenth centuries CE) (Reynolds 1955). The physical materials of primary interest to early scholars were antiques and artefacts collectable for their aesthetic appeal. Nevertheless, a systematic approach to archaeology that recognized the need to document ruins and topography was pioneered by the Italian historian Flavio Biondo (early fiftieth century CE), for which he has been called one of the first archaeologists (Reynolds 1955). Further advances took place in Britain, where clergymen began documenting details of the landscape and ancient monuments within their parishes (Taylor 1995). In the sixteenth and seventeenth centuries CE, antiquarians such as John Leland, John Aubrey and William Stukeley conducted surveys of the country, drawing, describing, and interpreting the monuments they encountered (Fox 1999). Archaeology remained largely speculative until it entered the excavation phase in the eighteenth century CE. Led by Thomas Jefferson (later the third American President), archaeology was now characterized by a scientific methodology for careful excavation followed by deductive and inductive reasoning (Wheeler 2004).

By the nineteenth century CE, archaeologists fully recognized the value of collecting and analysing all objects that may have been connected to past human settlements, regardless of their artistic appeal (Trigger 1994; Dever 1989). In the early twentieth century CE, the British archaeologist Mortimer Wheeler developed a highly disciplined system of stratigraphic excavation based on the simple principle of superimposition: older layers or strata are progressively buried by younger deposits (Piggott 1977). By recognizing that separate layers correspond to separate time ranges, archaeologists had a powerful and universally applicable technique to sequence developments at any given site (Higginbotham 1985).

Another idea with profound and universal implications for archaeology was put forward by the American anthropologist Julian Steward (twentieth century CE): everyday activities of humans in settlements—whether related to agriculture, or constructing buildings, or exploiting natural resources—constantly scar the earth's surface. Therefore, these scars are yet another form of physical evidence that archaeologists can leverage. A fellow American, the archaeologist Gordon Willey, was one of the earliest to demonstrate the power of Steward's idea by studying settlements in the Viru Valley, Peru (1946). Instead of examining individual settlements, Willey studied the valley as a whole using a combination of observations from aerial photographs (translated into detailed maps and checked in the field with compass and chain measurements), a ground-level survey, excavation and surface potsherds collection, and recording details of the setting and forms of architecture to establish dates for hundreds of identified sites (Vogt 2004). By plotting the geographical distribution of sites within the valley at different periods, Willey was able to establish associations with the changing local environment (Renfrew and Bahn 1996). This project emphasized that it is important for archaeologists to view sites holistically, and to account for environmental factors acting on past societies. Section 2.3.6 describes a similar study on the Sarasvati river system and demonstrates a connection between the distribution of Harappan sites and environmental factors.

1.2 Tools to Interrogate Spatial Contexts

At first, it is puzzling why archaeologists apparently overlooked traces of past human activities on the earth's surface as a source of evidence for so long. The answer is quite simple: many of these traces are spread across large area and are difficult (or impossible) to observe at ground level. It was only with the advent of aviation that archaeologists had access to an entirely new perspective: the synoptic view. Aerial and, more recently, space platforms are tools that give archaeologists an overall view of a site's spatial context.

As we shall see in Chap. 2, seemingly scattered features can sometimes be recognized as parts of overall patterns when viewed from an elevated platform (Trumpler 2005). This was first recognized by military reconnaissance pilots such as George Beazeley, a British Lieutenant-Colonel in the First World War. Beazeley spotted distinct outlines of ancient canals during his repeated flights over the Tigris–Euphrates plain in modern Iraq. He also discovered the remains of ancient Samarra about 130 km north-west of Baghdad. After the war, the British archaeologist Osbert Crawford explored the potential of wartime aerial photographs for archaeological research (Deuel 1969). By the Second World War, aerial photography was extensively applied for many archaeological purposes. In 1957, the British archaeologist John S.P. Bradford wrote the first book on interpreting aerial photographs and their implications for field archaeology (Bradford 1957).

This period coincides with the dawn of the space age and the Cold War. Once again, a host of technologies developed for military purposes during this time and in later decades found numerous applications in civilian domains, including archaeology. One of these was the rapid development of *remote sensing technologies*. Very quickly, aerial and satellite platforms were equipped not only with increasingly sophisticated photographic cameras, but with sensors to image in infrared and other longer wavelengths. In addition to these passive sensors, active sensors such as microwave and LiDAR were developed and flown. As we shall see in Chap. 3, remote sensing technologies allow archaeologists to search for evidence in non-invasive ways. The explosive growth in computing power over the last few decades has added further riches to the toolkit: powerful image processing techniques, 3D visualization and photogrammetry (which we shall explore in Chap. 3), Geographic Information Systems (GIS) for storing, retrieving, integrating, analysing and visualizing spatial data (see Chap. 4), global navigation satellite systems (GNSS) (such as GPS) which assists in correlating field studies with remote sensing imagery (see Chap. 4), and UAVs (unmanned aerial vehicles, or drones). The latter provide yet another aerial platform, and we discuss the relative advantages and disadvantages of such platforms over space platforms (i.e. satellites) for archaeological applications in the next section.

Before closing this section, we note that the tools listed above can be applied to archaeology in ways that go beyond the scope of this book. For example, we will assume that a site of archaeological interest has already been identified, and we will use these tools to examine this site's spatial context (for instance, see Nalanda: Sect. 5.1 and Talakadu: Sect. 5.4) or that we have a general idea of where the site should be and we study the vicinity to identify site's location (for instance Lalbagh palace at Srirangapatna: Sect. 5.3 and Patna: Sect. 4.3.4). However, even before such an investigation, GIS tools can be used to scan a much larger area and predict which sites could be of archaeological interest (Gillespie et al. 2016). Further, while we will use these tools to investigate the area around a site, archaeological investigations can also be conducted at specific locations within this area using non-invasive remote sensing technologies such as ground penetration radar (Malik et al. 2017).

1.3 Aerial Versus Space Platforms

A platform is simply the stand on which a sensor is placed. Starting from ground level, platforms for archaeological applications can place sensors at various altitudes, ranging from a few metres to tens of metres (towers), tens to mid hundreds of metres (kites), tens to high hundreds of metres (UAVs), hundreds of metres to a few kilometres (aircraft), hundreds of metres to tens of kilometres (balloons), or hundreds of kilometres (low-earth orbit satellites). Altitude and other characteristics (such as restrictions on payload and usage near archaeological sites) determine the perspective, area of coverage, and the cost for images taken from each platform (Fig. 1.2). Table 1.1 summarizes these advantages and disadvantages.

It is clear from this table that certain platforms will be more appropriate than others for studying specific sites. For instance, since remote sensing satellites orbit at an altitude of about 600–900 km, they may be unable to provide images of sufficient quality (e.g. due to atmospheric distortion) for certain analyses, or may not support certain types of sensors (e.g. LiDAR). On the other hand, since these satellites orbit earth in a near-polar sun-synchronous orbit, they are the only platform that can cost-effectively target every location on earth and can repeatedly image that location under similar solar illumination. For these reasons, the examples and case studies presented in this book will primarily use images taken from space platforms.

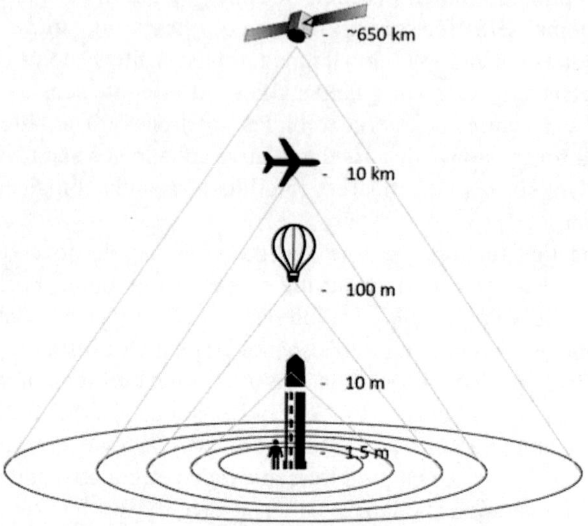

Fig. 1.2 Synoptic view and coverage area based on altitude of platform

Table 1.1 Advantages and disadvantages of observations from various platforms

Platform	Advantages	Disadvantages
Tower	Heavy payload	Single oblique perspective (potentially exacerbated by site restrictions), high transport and set-up cost
Kite	Oblique and vertical perspectives (unless there are site restrictions)	Very light payload, adverse wind conditions can impact operational cost and stability
UAV	Oblique and vertical perspectives (unless there are site restrictions)	Light payload, high operational cost, adverse wind conditions can impact operational cost
Aircraft	Oblique and vertical perspectives (unless there are site restrictions), very heavy payload	Very high operational cost
Balloon	Oblique and vertical perspectives, heavy payload	Medium operational cost which can increase due to adverse wind conditions
Satellite	Multiday vertical perspectives under similar lighting conditions, heavy payload, low operational cost (if amortized across multiple applications)	Very high set-up cost, atmospheric interference

1.4 How This Book Is Organized

To successfully apply these technologies to archaeology, it is necessary to understand how they work and to appreciate their limitations. To assist the reader, this book is organized as follows:

Chapter 2 describes how landscapes evolve over time and discusses the implications of this evolution for the analyst seeking to identify archaeological features from above. Thus, this chapter provides the basis for answering critical questions such as "What kinds of features should I be looking for in satellite images?" and "Why can't I see what I'm looking for in this image?"

Chapter 3 describes the various types of imaging technologies that are operational on contemporary and retired earth observation satellites. This chapter helps the analyst answer the question "Is this type of satellite image likely to be useful for my analysis?". The chapter also reveals some of the science behind these technologies to help answer the inevitable follow-up questions "Why?" and "Why not?".

Chapter 4 answers the question "How is GIS used for archaeological investigations?" Our presentation is deliberately limited to this application domain, and at a scale that covers the spatial context of a site, to avoid overwhelming readers who are unfamiliar with GIS with complexities that a general introduction to GIS must necessarily address.

Chapter 5 presents five case studies to demonstrate that there is no fixed answer to the question "How should I conduct geospatial analysis for sites of interest?" This chapter also reemphasizes the core ideas presented in Chaps. 2–4 in more complex settings.

Chapter 6 explores the question "How can geospatial analysis be used for preserving, conserving, and managing cultural heritage sites?". It examines some of the limitations of current approaches and proposes alternatives that address these limitations.

Finally, Chap. 7 provides an affirmative answer to the question "Can geospatial analysis of archaeological sites contribute to multidisciplinary research?" through three case studies. It also discusses future directions for this body of research.

References

Bradford J (1957) Ancient landscapes: studies in field archaeology. G. Bell and Sons Ltd, London
Deuel L (1969) Flights into yesterday: the story of aerial archaeology. Pelican, Middlesex
Dever W (1989) The Biblical Archaeologist 52(2/3):146. https://doi.org/10.2307/3210206
Fox A (1999) Remembering the past in early modern England: oral and written tradition. Trans R Hist Soc 6(9):233–256
Gillespie TW, Smith ML, Barron S, Kalra K, Rovzar C (2016) Predictive modelling for archaeological sites: Ashokan edicts from the Indian subcontinent. Curr Sci 110(10) 25 May:1916–1921
Herodotus (2006) The histories (trans: Aubrey de Sélincourt 1954) Introduction-John Marincola. The Folio Society, London
Higginbotham E (1985) Excavation techniques in historical archaeology. Aus Hist Archaeol 3:8–14

Malik JN, Gadhavi MS, Satuluri S, Kumar S, Sahoo S, Vikramam B (2017) Unravelling the hidden truth from Vigukot in the Great Rann of Kachchh, western India by surface and sub-surface mapping. Curr Sci 113(10) 25 Nov:1906–1917

Piggott S (1977) Robert Eric Mortimer Wheeler: 10 September 1890–22 July 1976. Biogr Mem Fellows R Soc 23(Nov):623–642

Renfrew C, Bahn P (1996) Archaeology: theories, methods and practice, 2nd edn. Thames & Hudson, London, p 17

Reynolds BR (1955) Latin historiography: a survey, 1400-1600. Stud Renaissance 2:7–66

Taylor B (1995) Amateurs, professionals and the knowledge of archaeology. Br J Sociol 46(3) Sep:499–508

Trigger BG (1994) Pitt Rivers: the life and archaeological work of Lieutenant-General Augustus Henry Lane Fox Pitt Rivers, DCL, FRS, FSA by Mark Bowden. Am Antiq 59(2) Apr:390–391

Trumpler C (ed) (2005) The past form above. Francis Lincoln Limited, London

Vogt EJ (2004) Gordon Randolph Willey. Bibliographical memoirs, vol 84. The National Academies Press, Washington DC

Wheeler M (2004) Archaeology from the earth. Munshiram Manoharlal Publishers Pvt Ltd., Delhi (1st pub 1954)

Chapter 2
Landscape Morphology and Spatial Patterning of Archaeological Signatures When Viewed from Above

Preamble

We will refer to the landscape in the vicinity of a site as its *spatial context*. For a given site, we would like to identify features in satellite images of its spatial context that may provide new archaeological insights for that site. Images may directly show certain surface features, while some subsurface features may only be visible indirectly through certain morphological expressions. To help readers recognize features of potential interest, this chapter will examine several examples of such landscape features in synoptic views. (In subsequent chapters, we will see how these features in satellite images can sometimes be enhanced using a variety of sensors, or through image processing techniques). We will also explain why these features are not always visible in satellite images. This will require a complex argument, but the key reason is because a site's spatial context is not constant—it undergoes continuous evolution due to a combination of natural and human factors which depends significantly on the type of landcover. With this understanding, we will appreciate why no single approach to identify such features is likely to work at all sites.

2.1 Gradual Versus Rapid Change

A site's spatial context is continually undergoing gradual change, caused by a combination of *natural factors* (e.g. when a structure is weathered by wind or water, or buried by the deposition of sand or silt, or covered by vegetation) and *anthropogenic* or *human factors* originating in human activity (e.g. opportunistically mining material for reuse). Figure 2.1 illustrates gradual change to a deserted fort and moat caused by gradual natural and human factors. Rapid change can also occur sporadically, once again due either to natural factors (e.g. natural disasters) or human factors (e.g. large-scale redevelopment or conservation efforts). We will briefly discuss the

M. B. Rajani, *Patterns in Past Settlements: Geospatial Analysis of Imprints of Cultural Heritage on Landscapes*, Springer Remote Sensing/Photogrammetry,
https://doi.org/10.1007/978-981-15-7466-5_2

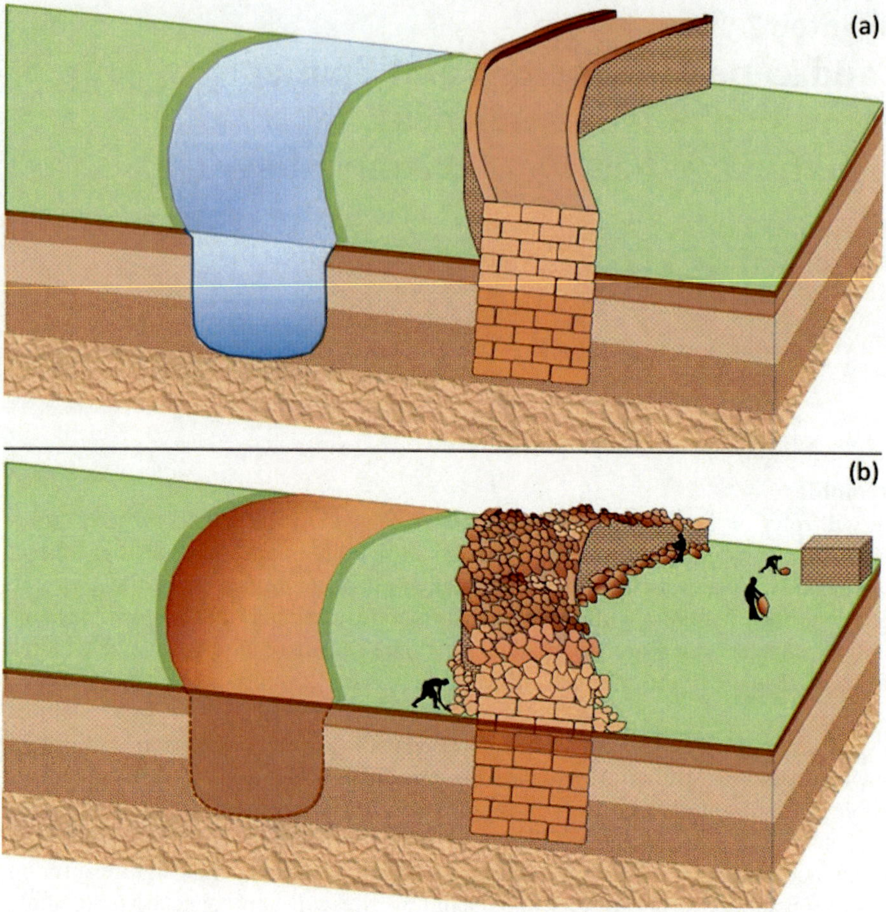

Fig. 2.1 Diagram showing **a** a typical fort structure with an adjoining moat; **b** the same after a period of gradual deterioration

former type in the following subsection and the latter (which can be mitigated by using geospatial technologies) in Chap. 6. However, the bulk of this chapter will focus on gradual change, primarily because it occurs all the time at every site.

2.1.1 Rapid Changes Due to Disasters

Landcover can change suddenly and drastically because of natural and anthropogenic disasters. When such events occurred before the availability of aerial views, their effects were not easy to detect. Today, we usually have images from before and after the event. Comparing these images can be extremely helpful in assessing the extent of the damage and in planning mitigation. For example, the mud-built citadel of Bam in Iran crumbled due to an earthquake on 26 December 2003, which measured 6.6 on the Richter scale. Satellite imagery and aerial views were used to assess the overall damage that the site incurred (Rouhi 2016). Precisely one year later, a Tsunami triggered by an earthquake in the Indian Ocean had a devastating effect on life and property along the coasts of Indonesia, Sri Lanka, India, and several other countries. In India, the force of the receding waves also washed away the sediments on the shore and uncovered several objects of archaeological interest. One of these was a tenth-century inscription engraved on a boulder at Saluvankuppam (6 km north of Mamallapuram) indicating the existence of a Subramanya temple. A subsequent excavation exposed a whole temple complex (Bhadreenath et al. 2011). Figure 2.2 shows the site before Tsunami and after excavation.

Fig. 2.2 Saluvankuppam **a** before the December 2004 Tsunami; **b** showing the Subramania temple after excavation

2.1.2 Gradual Changes

When we look at a site's spatial context today, we see only the present point in its evolutionary trajectory. For some sites, we may have historical spatial records that provide multiple snapshots of these spatial contexts as they change over time. These records may date from a few years ago (e.g. Google Earth images) to a few decades ago (e.g. Corona and other satellite imagery) to a few centuries ago (e.g. old maps, paintings, and spatial descriptions in historical texts). Even with multiple snapshots for a site, our data is generally too sparse to precisely infer the sequence of gradual changes to its spatial context. This lack of data forces us to guess how the spatial context evolved and to use these guesses to explain why we can or cannot observe certain features. While this may seem highly unscientific, it is standard practice in science to explain observations based on a minimal number of plausible guesses (or *hypotheses*), and to be proved wrong if contradictory evidence is found.

Formulating these guesses becomes quite intuitive with practice. To help build this intuition, we split our discussion of gradual change into multiple sections, each of which considers different dimensions along which features can vary. We note that the insights presented in these sections are based on our experience with a diverse but limited set of sites. It is quite possible that further experience will grow this collection of insights and add nuance to the discussions that follow.

2.2 Indirect Versus Direct Evidence

Certain large-scale features provide indirect evidence for past settlements, i.e. the presence of such features does not necessarily imply the presence of nearby settlements. These features, which include palaeochannels (past rivers or streams that are now inactive), mudflats, and coastal strandlines (features which may indicate past coastlines) typically span multiple kilometres. Because of their size, they are sometimes difficult to see amid the extraneous clutter of small-scale features in high-resolution satellite images. Instead, large-scale features are often easier to identify in medium-resolution images (~20–30 m per pixel). Figure 2.3a shows the palaeochannel in northern Rajasthan along which many Harappan sites lie (see also Sect. 4.1). The meandering pattern of the erstwhile river is visible as a darker tone in the Landsat image (30 m per pixel). The extreme top-right of this image (the spatial context of the site Kalibangan) is shown at a higher resolution in Fig. 2.3b. Notice that the shape of the palaeochannel is inconspicuous amid the details of parcel boundaries, roads, settlements, etc.

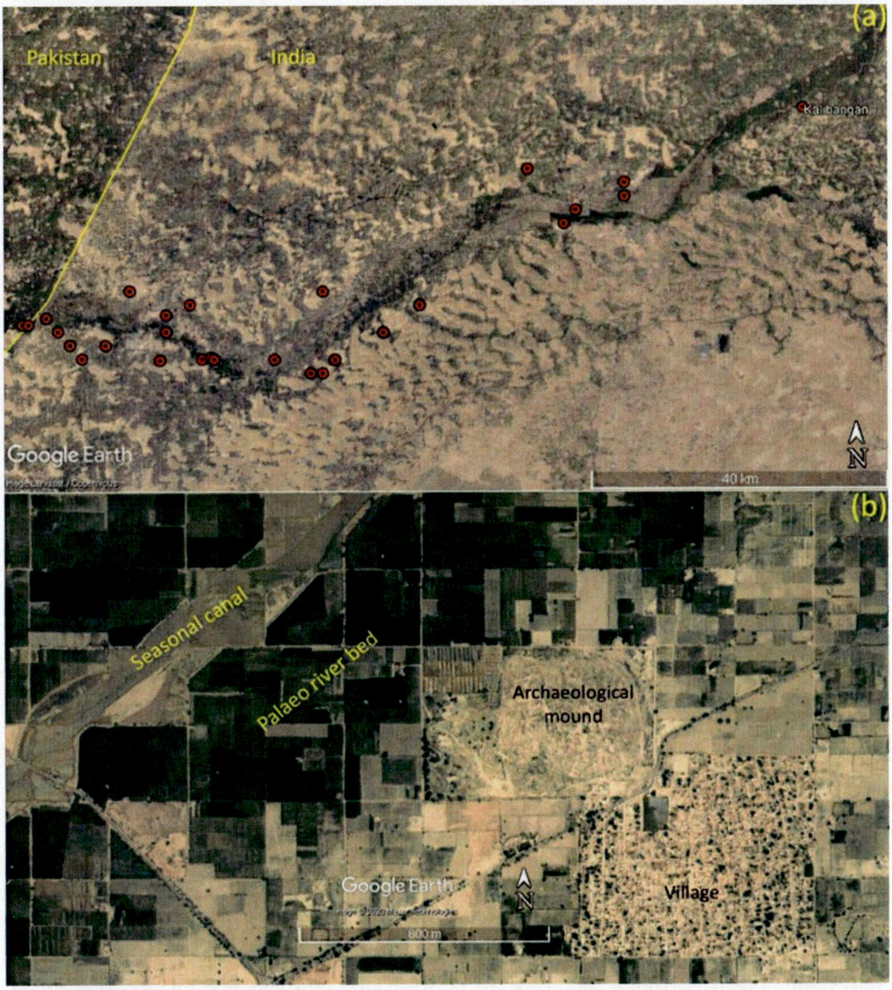

Fig. 2.3 **a** Harappan sites dotted along the seasonal stream and palaeochannels of Ghaggar-Hakra in northern Rajasthan (the site Kalibangan is on top-right corner), Landsat/Copernicus; **b** Kalibangan and environs seen on high-res image (GE Maxar Technologies)

Man-made features such as buildings (whole or partial), ditches, pits, canals, moats, tanks, and ponds constitute direct evidence for past settlements. Even when none of their remains seem visible at ground level, these features can leave traces that are sometimes seen in satellite images in the form of cropmarks, soil marks, field boundaries, or urban land-use boundaries. These comparatively small-scale features typically span a few tens to a few hundreds of metres and are best identified in high-resolution images (5 m or less per pixel).

2.3 Differences in Landcover

A site's spatial context often has one predominant type of landcover. The type of landcover often limits the set of plausible changes that could have occurred within that spatial context. We therefore consider several typical types of landcover in the following subsections. We note that some spatial contexts may have multiple types of landcover.

2.3.1 *Arid Soil*

Arid lands have little or no plant cover and are often sparsely inhabited. Thus, the landcover is characterized by bare soil that is largely uniform in colour. However, even a seemingly homogeneous land parcel can show slight variations in colour, caused by differences in mineral and organic content of the soil and its moisture content. When these colour variations appear as anthropogenic patterns (Fig. 2.4), they suggest traces of past settlements. Thakker (2001) has demonstrated the value of identifying such patterns in revealing the existence of archaeological sites in parts of Kutch (Gujarat). Soil marks can also be visible in non-arid or comparatively wet and fertile land when the soil is left uncultivated.

Buried structures can also influence the colour, tone, and texture of surface soil, depending on how much residual building material is on the surface. Figure 2.4 shows four examples. When the surface material is not fully removed and gets covered over time, it can produce a ridge (Fig. 2.4a) or a series of intermittent mounds (Fig. 2.4b). Even when the structure is entirely buried, variations in the colour or tone of the soil due to differences in moisture content (Fig. 2.4c) or vegetation cover (Fig. 2.4d) can reveal patterns that indicate the underlying structure. Figure 2.5 shows the rectangular external wall and foundations as well as some internal walls of the archaeological settlement at Dholavira (an example of Fig. 2.4d). This cropmark is indicated by Babul trees that grow like a weed in the region. This site was excavated from 1989 to 2005, but the Corona image shows these cropmarks prior to their excavation. The buried remains of a fortification at Ahichhatra, U.P. (Fig. 2.6), is an example of the patterns illustrated in Fig. 2.4a, b.

2.3.2 *Agricultural and Semi-agricultural Land*

We use the term *semi-agricultural* when the predominant landcover is agricultural together with small settlements and water bodies. For agricultural land, buried archaeological remains often affect the health of crops, creating positive or negative cropmarks which reveal themselves as large patterns when viewed synoptically (Bradford 1957; Wilson 2000). Cropmarks are not only one of the most common signals for

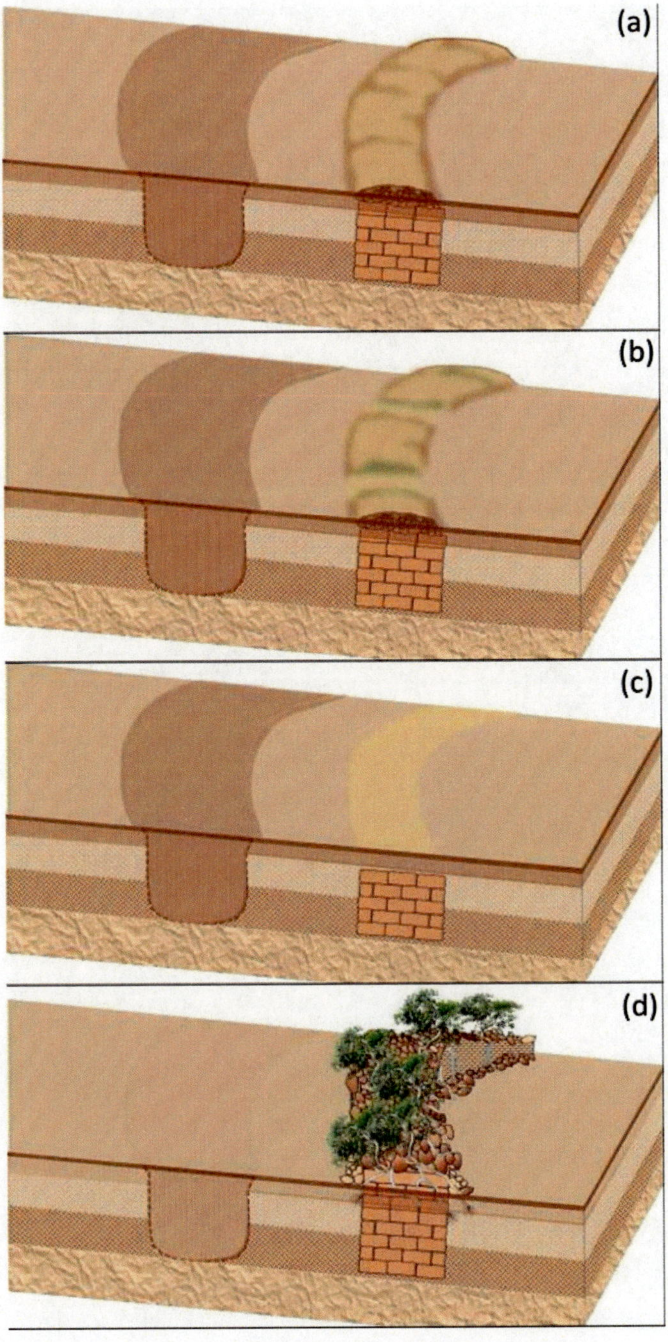

Fig. 2.4 Diagram showing typical patterns observed in arid soil indicating buried heritage structures

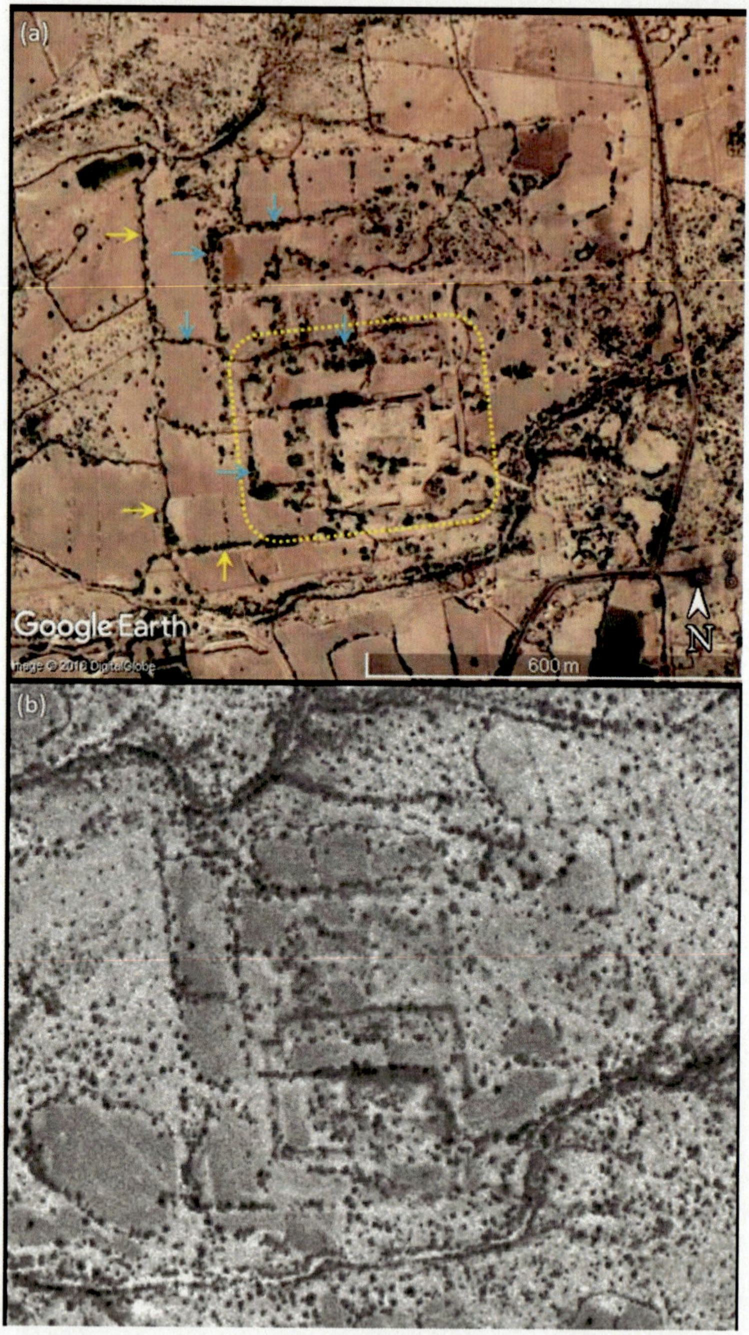

Fig. 2.5 Dholavira **a** as seen on 31 December 2016 (yellow arrows indicate external walls, cyan arrows indicate internal walls, and the yellow dotted line marks the excavated region); **b** Corona image taken on 31 December 1965

Fig. 2.6 Buried remains of fortification at Ahichhatra, Uttar Pradesh

detecting archaeological remains, but also one of the oldest—they were seen in aerial reconnaissance photographs taken in World War II and recognized for their archaeological importance (Trumpler 2005). Cropmarks can reveal disused moats, canals, tanks, and pits. Since these features (when buried and silted) often hold additional moisture, they typically appear as positive cropmarks in agricultural land (Fig. 2.7a). In contrast, when archaeological structures such as brick/stone walled foundations, streets and solid floors are buried beneath soil, they tend to inhibit the growth of vegetation because they obstruct plant roots (Fig. 2.7a and Fig. Box 1), and hence, they appear as negative cropmarks.

Since most crops have an annual cycle, cropmarks need not be visible in all seasons. Depending on the depth to which roots penetrate and the depth of archaeological remains, it is possible that cropmarks are only seen during extreme weather conditions such as peak summer, when moisture and nutrients in the upper layers are exhausted and roots must penetrate deeper. Further, they may only be visible when viewed in certain wavelengths (see Sect. 3.2.1). Hence, it is advisable to analyse images from multiple seasons using a variety of sensors to identify such features with greater confidence.

Neat and well-defined cropmarks as illustrated in Fig. 2.7a get created when the subsurface composition is nearly uniform across the feature. If subsurface material in the foundation of a structure was partially mined for reuse, there may be irregular cavities that hold moisture. In such cases, we might see an irregular mixture of positive and negative cropmarks (Fig. 2.7b). Generally, linear features (e.g. caused by fort walls and moats) are easier to detect than nonlinear features (e.g. caused by buried remains of buildings). Depending on how much building material has been

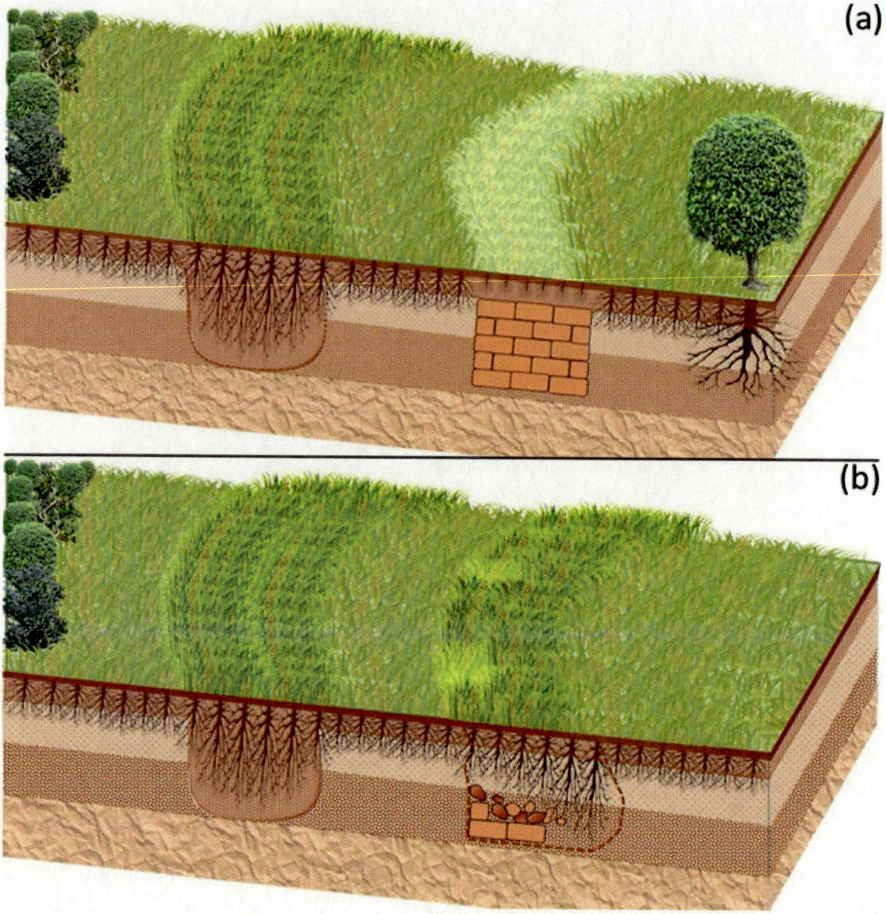

Fig. 2.7 Diagram showing **a** positive and negative cropmarks in agricultural land; **b** irregular mixture of positive and negative cropmarks when subsurface material has been partially mined

removed from the subsurface foundation, the cropmark can be sharper or obscure. If the surface building material is not fully removed, then the remains of separate structures can form individual mounds.

Former water bodies that subsequently became agricultural land also form non-linear features. Figure 2.8 illustrates an example of a cluster of temples (labelled as 3,4,5,6) and water bodies (labelled as 1 and 2). Figure 2.8a illustrates the condition when these were active and Fig. 2.8b shows the locations of structures and water bodies covered by agricultural vegetation. Figure 2.8c reflects the same condition as Fig. 2.8b, but it is visualized as a false-colour image using the infrared band (Sect. 3.2.1) and shows both positive and negative cropmarks in starker contrast.

The landcover within a single parcel of land tends to be homogenous, which causes variations such as those shown in Fig. 2.7a, b to stand out. Another possibility

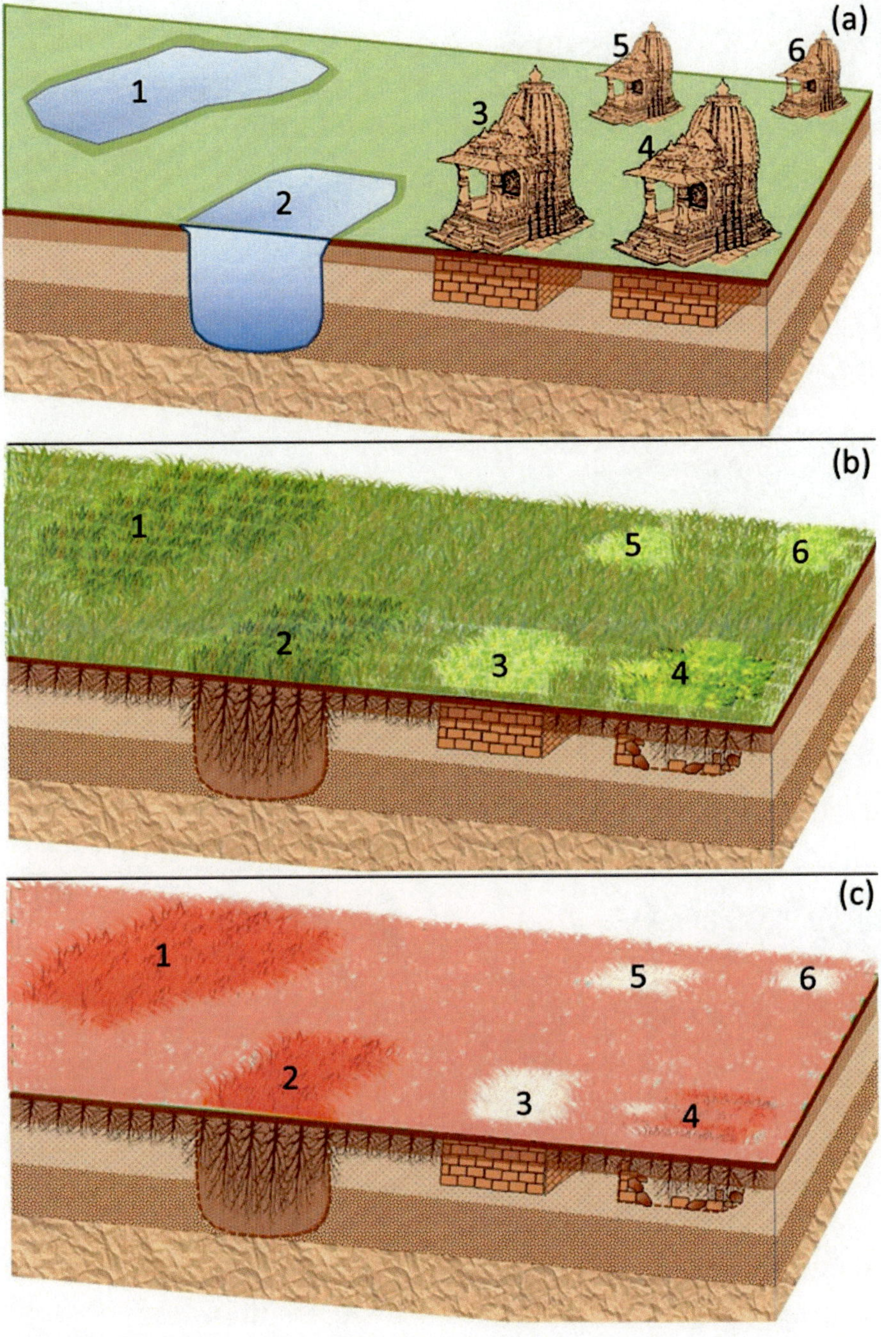

Fig. 2.8 Diagram showing **a** a site with temple structures and water bodies; **b** positive, negative, and mixed cropmarks indicating buried remains; **c** enhanced contrast between cropmarks as they might appear in infrared images

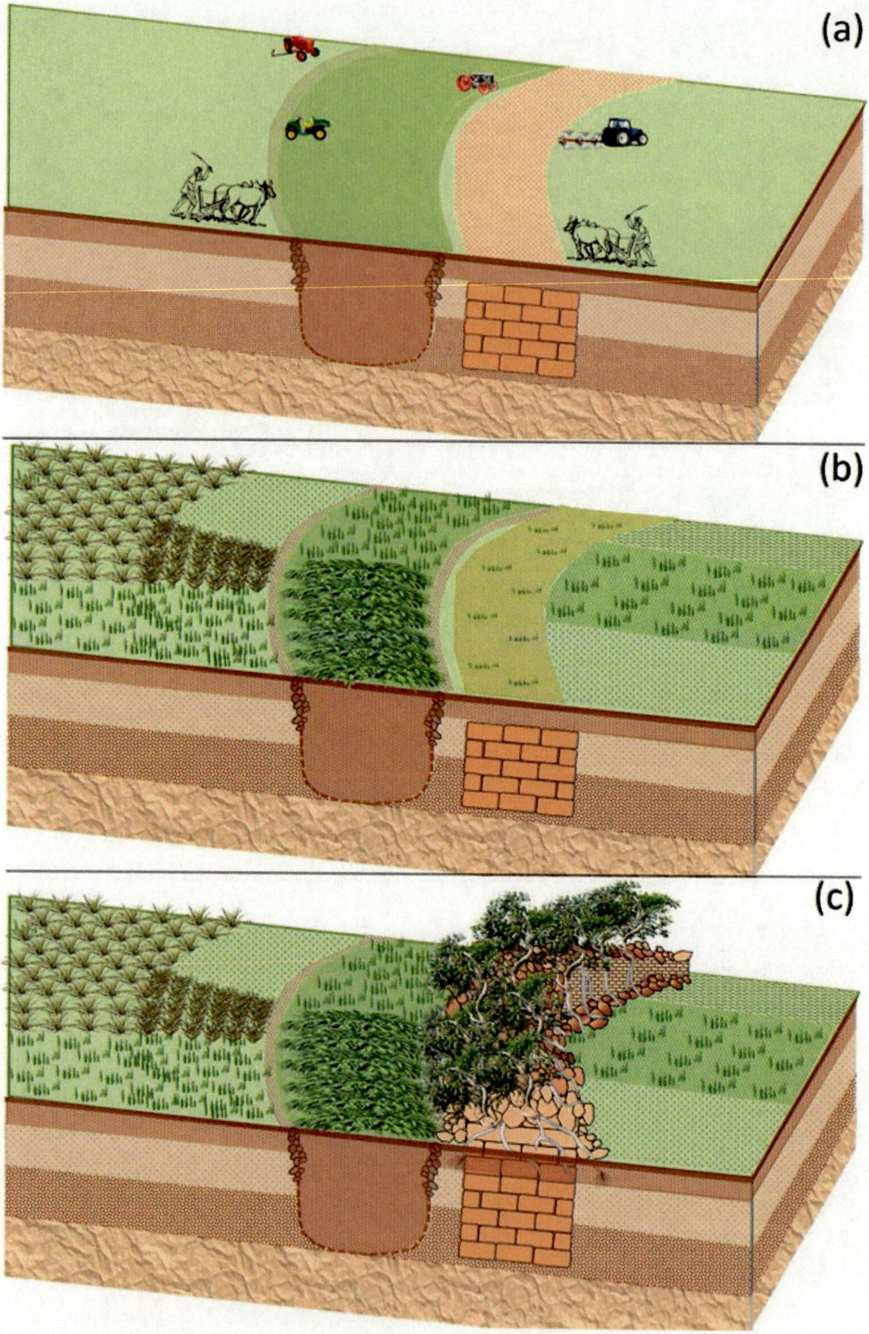

Fig. 2.9 Diagram showing buried structures revealed by a sequence of field boundaries

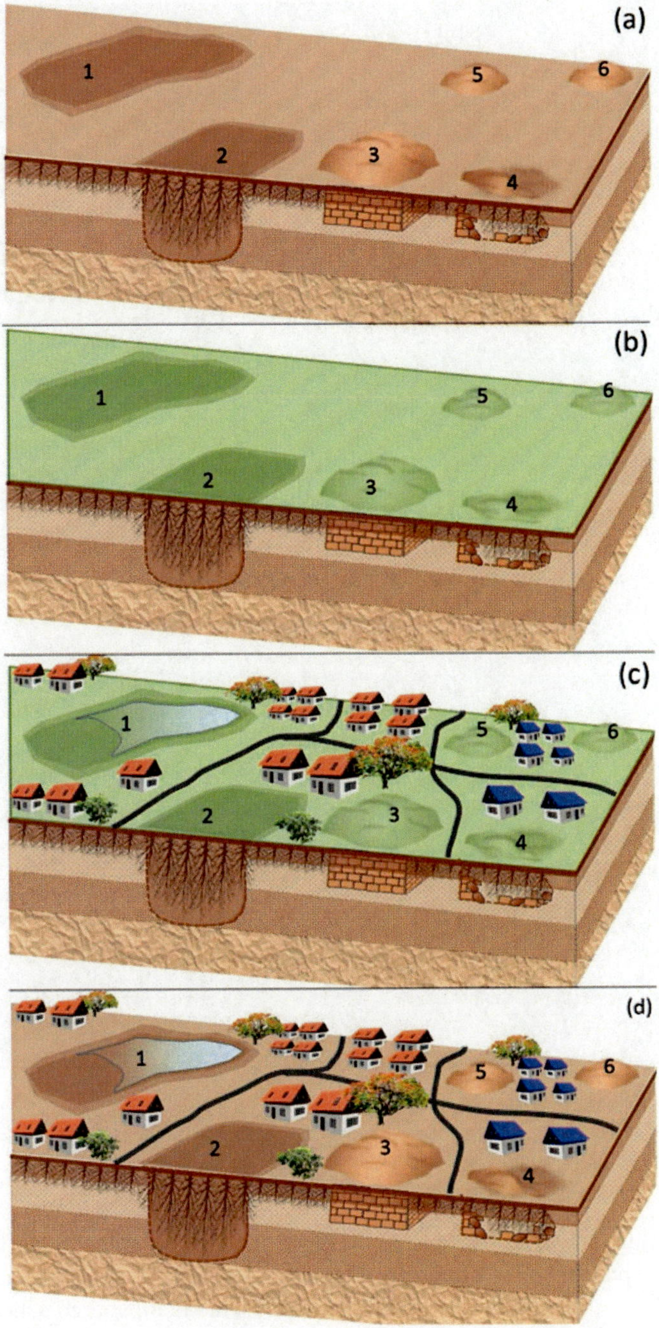

Fig. 2.10 Diagram showing mounds amidst settlement

Fig. 2.11 Halebidu, the fortified capital of the Hoysala dynasty. **a** Continuous multiple agricultural field boundaries indicating the shape of a past moat north of the fortified area; **b** thick vegetation on remains of a fort to the south-west; inset a field photograph

is that subsurface material causes enough hindrance to agriculture that it dictates the division of parcels. When this happens, a sequence of field boundaries forms a collective pattern that reveals the buried structure. This is illustrated schematically in Fig. 2.9. Individual buried structures may initially appear as distinct mounds even as the landcover undergoes changes due to natural factors (Fig. 2.10a, b). However,

Fig. 2.12 **a** Settlement of Baragaon and Surajpur together with its environs, north of the excavated site of Nalanda; **b** close up of the unexcavated mound in Baragaon

when humans build subsequent settlements, these distinct mounds may amalgamate (Fig. 2.10c, d).

A large settlement may lie beneath multiple types of landcover. By way of illustration, we now look at two such settlements that exemplify many of the changes we have discussed above. First, consider the fortified settlement of Halebidu, the capital of the Hoysala dynasty which ruled much of what is now Karnataka in the

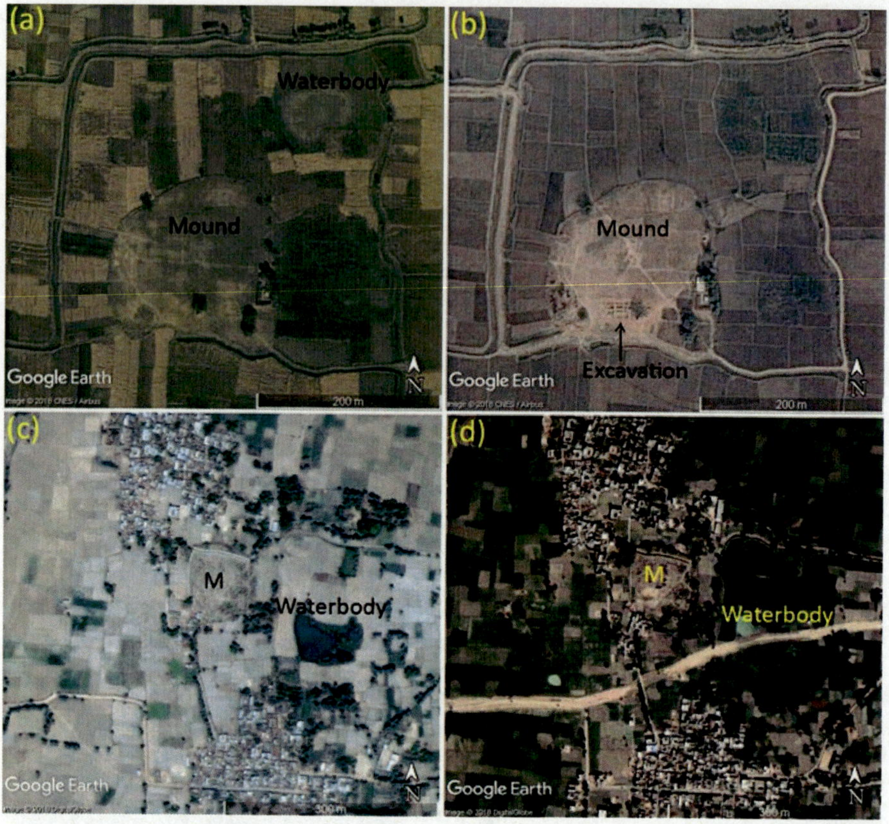

Fig. 2.13 Archaeological mound of Jagdispur, 2 km south-west of the site of Nalanda: **a** 20 November 2006, **b** 4 December 2015; mound (M) in the settlement of Jeofardih 1.5 km west of Nalanda: **c** 8 May 2010, **d** 7 March 2018

twelfth century CE. The landcover is predominantly agricultural—except for the settlement at the northeast corner. The northern portion shows a sequence of several land parcels that reveal the curvilinear shape of the moat (Fig. 2.11a), even though almost no structural material of the fort survives here. In contrast, ruins of the fort made of huge stone blocks survive to the south and west. On the south-west, the remains of the fort lie beneath thick vegetation that is distinct from the vegetation in adjacent agricultural fields (Fig. 2.11b). This vegetation reveals the shape of the fort, including its bastions, because the plants here appear to have found ample moisture and nutrients within cavities (as illustrated in Fig. 2.9c) (Rajani and Kasturirangan 2014, Das and Rajanai 2020).

As a second example, consider the spatial context of Nalanda. Figure 2.12 shows the area north of the excavated site of Nalanda with an active tank and two positive cropmarks indicating past water bodies that are now used for cultivation. In addition, we see a settlement (Baragaon), an unexcavated mound, and an excavated structure (Temple 14). Figure 2.13 shows another mound in this spatial context (Jagdispur;

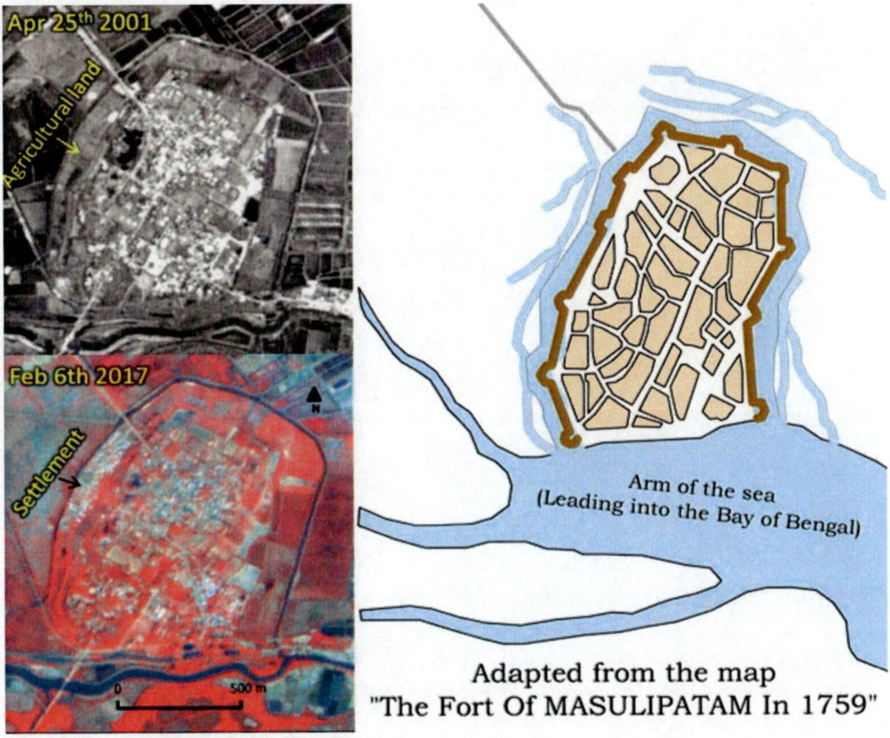

Fig. 2.14 Gradual change in the landcover of individual parcels at Masulipatnam over time

1.5 km south-west of excavated site). It is located on agricultural land and is distinctly visible from the surrounding. The two images (a) and (b) are taken on different dates—a water body on the northeast side of the mound is clearly visible in Fig. 2.13a, but not in Fig. 2.13b. The most probable explanation for this is that the water body is depressed relative to its surrounding area—it gets flooded in the rainy season and is used for cultivation at other times. South of this mound (Fig. 2.13b), a checkered pattern of excavation can be seen. Figure 2.13c, d are of Jeofardih, which is amidst both agriculture and settlement. Its shape is therefore more obscure than Jagdispur in a satellite image. Jeofardih (1.3 km west of excavated site of Nalanda) also has an adjacent tank and Fig. 2.13d shows the construction of a new road that cuts the tank into two halves. This may cause the tank to dry up completely.

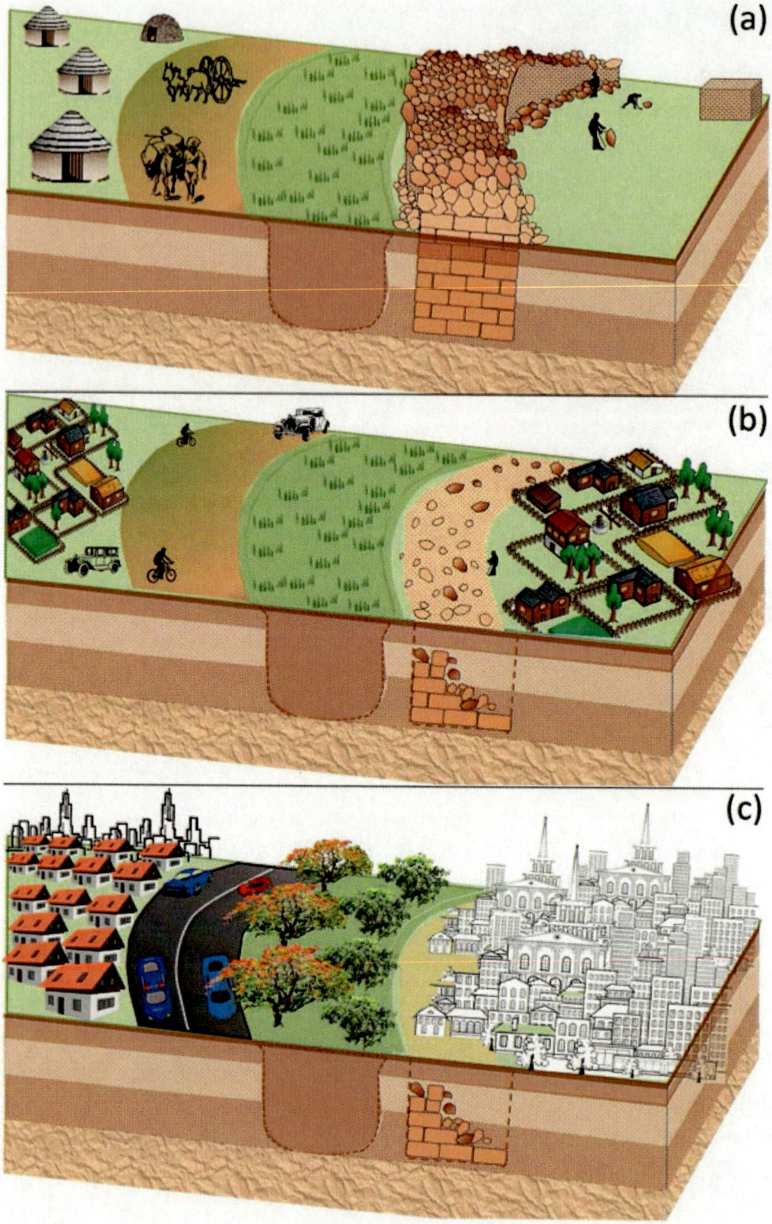

Fig. 2.15 Diagram showing morphology in urban landscapes

Box 1: It is critical to understand key properties of soil, because it is the medium through which the presence of subsurface archaeological remains are communicated before they are revealed by surface vegetation. Soil is biologically active and porous and has developed in the uppermost layer of the Earth's crust. Soil is composed of distinct layers called horizons that run roughly parallel to the surface (Fig. Box 1).

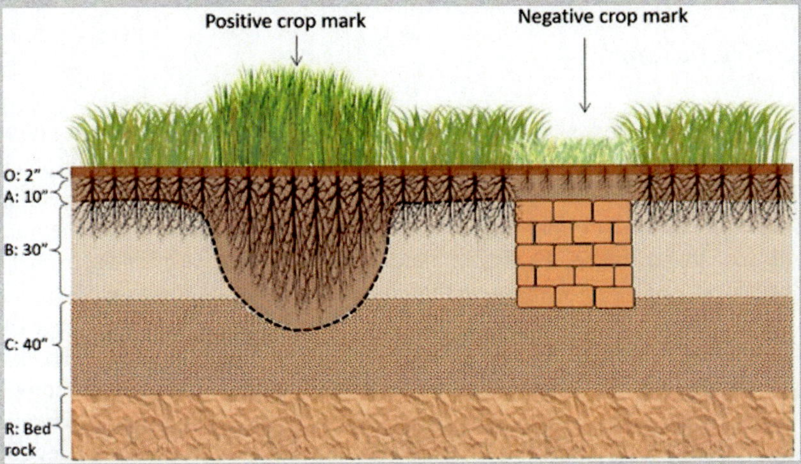

Fig. Box 1 Diagram showing positive and negative cropmarks over buried archaeological features

Each horizon has different properties and characteristics. A soil profile is a vertical section that extends from the surface to the underlying rock material. The surface horizon or O-horizon is comprised of organic material in various stages of decomposition. It is most prominent in forested areas where there is accumulation of debris fallen from trees. The A-horizon lies below the O-horizon and largely consists of minerals (sand, silt and clay) mixed with considerable amounts of organic matter and soil life. Below this lies the B-horizon, which is a site of deposition of certain minerals that have leached from the layer(s) above. The C-horizon is the least weathered and is comprised of large pieces of loose rocks. Finally, the R-horizon is the bedrock underlying the soil, which largely comprised of continuous masses of hard rock that cannot be excavated by hand (Soil Profile 2007–2020). These horizons vary in thickness and colour.[1] When a moat was excavated in the past, a slice of these horizons would have been removed. The subsequent silt deposition would have occupied this cavity pushing the lower limit of the A-horizon further down and creating a large space for minerals, organic matter, and soil life. This would allow the roots of surface vegetation to penetrate further and avail more nutrients. As a result, these crops tend to be healthier and denser, creating positive cropmarks indicating the moat (Fig. Box 1). In contrast, buried walls prevent roots from penetrating deep below the surface, allowing only a thin O-horizon. Therefore, the surface vegetation tends to be stunted and sparse, forming negative cropmarks (Fig. Box 1).

2.3.3 *Urban Land*

Many modern cities have expanded from past fortified settlements. Unless there is disruptive change to the layout within the fortifications, we often see gradual changes to the landcover for individual parcels of land (or adjacent groups of land parcels). As an example of this kind of landscape morphology, compare the two satellite images of Masulipatnam in Andhra Pradesh (taken 16 years apart) with a map from 1759 CE (Beveridge 1900), showing how the erstwhile layout of the fort has mutated into individual land parcels (Fig. 2.14). It is clear from the satellite images that the landcover in these parcels has gradually changed. For instance, a parcel that was agricultural in 2001 was developed into a settlement by 2017. Archaeological settlements that now lie in urban landscapes typically change in a similar parcel-by-parcel change and apart from a few sacred structures, most structures are rebuilt or extensively modified over time. This period can vary from site to site.

In contrast to structures, arterial roads tend to remain intact and the shapes of fortified boundaries (past fort walls/moats) are usually preserved in the form of additional roads. The diagram in Fig. 2.15 illustrates a plausible chain of events. Figure 2.15a shows a dilapidated fort wall. Having become a hindrance for mobility, people and goods move along the hindrance rather than cutting through it, creating a path. Eventually, the path becomes road (Fig. 2.15b). In due course, the ruins of the fort that were once a hindrance are fully cleared, and the land use on either side of the road continues to evolve. By this time, however, the road has gained enough importance to serve as a permanent marker of the shape of the past fort/moat (Fig. 2.15c). An excellent example of such a change is seen in the transformation of urban landcover east of Qila Rai Pithora over a half-century period (Fig. 3.13a versus Fig. 3.13b).

[1] https://www.ctahr.hawaii.edu/mauisoil/a_profile.aspx. Accessed 12 Apr 2020.

Fig. 2.16 Srirangapatna **a** yellow, cyan, and orange arrows point to cropmarks indicating three concentric moats (outer, middle, and inner, respectively) carved deep into the bedrock; **b** ground photograph showing a bastion built over the rock

As an example, the highway SH17 in Srirangapatna skirts the fort (Fig. 2.16). Similarly, several fortified settlements depicted in maps of Old Delhi/Shahjahanabad[2] (Fig. 2.17), Ahmedabad[3] (Fig. 2.18), Madurai[4] (Fig. 2.19) and Bombay Fort[5] (Fig. 2.20) are likely to have undergone similar transformations. In each of these cases, one can see the internal road still in existence, and more modern roads

[2] http://www.columbia.edu/itc/mealac/pritchett/00routesdata/1800_1899/ghalib/delhimap/delhimap.html Accessed 07 May 2020.

[3] https://commons.wikimedia.org/wiki/File:Ahmedabad_City_and_Environ_Map_1866.jpg Accessed 07 May 2020.

[4] https://ssubbanna.files.wordpress.com/2012/09/565cd-ma28city-map_1380140g-madurai-sepia.jpg. Accessed 07 May 2020.

[5] https://commons.wikimedia.org/wiki/File:Bombay_Fort_1771-1864.jpg. Accessed 07 May 2020.

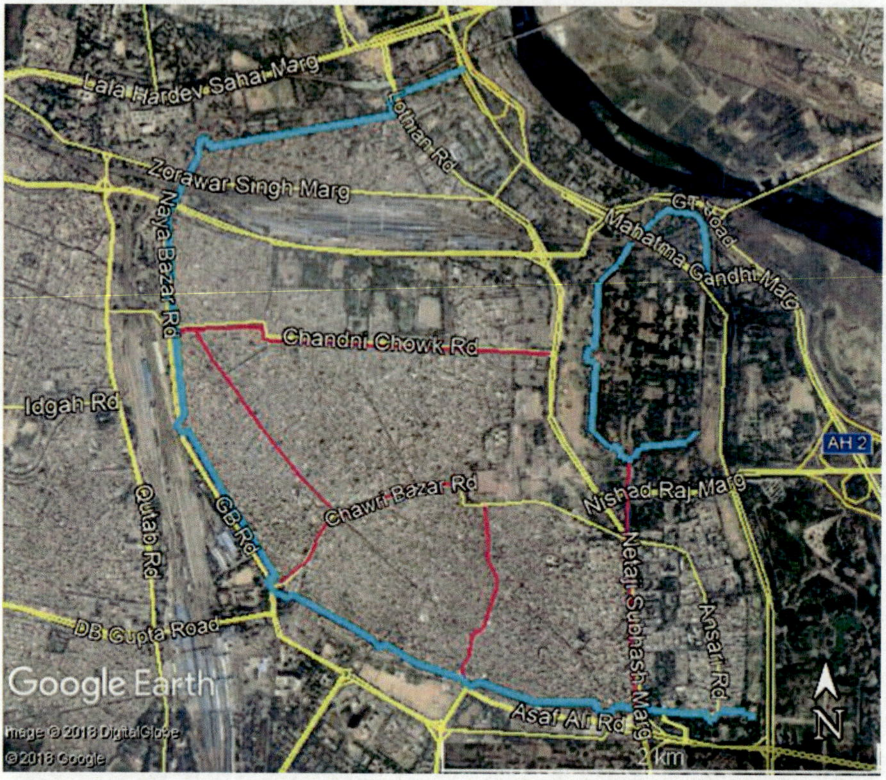

Fig. 2.17 Fort feature from *Plan of Delhi 1857–58* overlaid on Google Earth (fort walls are marked in cyan and major roads are marked in magenta)

following the profiles of past forts. This morphology is not unique to Indian settlements, of course. We see a similar change in Vienna, where the shape of the fort and moat marked in historical maps of Vienna (including the esplanade) can be matched with the settlement and road patterns in the current layout of central part of the city.[6] The core settlement of old Bangalore also illustrates this phenomenon (Fig. 2.21) (Rajani 2007). The situation here is slightly different, because Bangalore's fort had two distinct components: the *kote* (the royal enclosure shaped like an oval pendant) and the *pete* (the rest of settlement to the north of the *kote*). Whereas the *pete*'s shape is preserved by the roads surrounding it as described above (particularly when visualized with false-colour satellite images discussed in Sect. 3.2.1), the shape of the *kote* is unrecognizable in the road pattern. This is probably because of the disruptive changes that began in the nineteenth century CE when the fort was dismantled in

[6]Georeferenced overlays of historical maps of many parts of Europe can be visualized in: https://mapire.eu/en/map/europe-19century-secondsurvey/?bbox=1817747.7219660091%2C6139735.8597633075%2C1829815.217806531%2C6143557.711177567&map-list=1&layers=158%2C164. Accessed 07 May 2020.

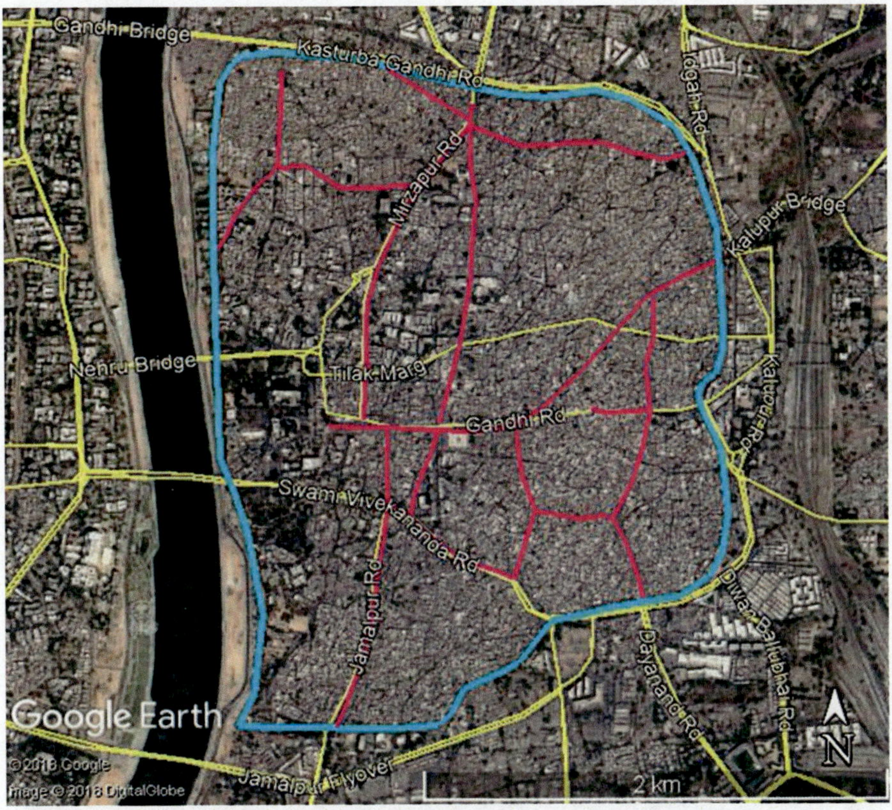

Fig. 2.18 Outline of the fort marked in the map *The city of Ahmedabad with its environs* (1866) overlaid on Google Earth (fort walls are marked in cyan and major roads are in magenta)

sections to make way for a fresh layout consisting of roads, colleges, schools, bus stands, and hospitals (Iyer 2019).

Srirangapatna, referred to earlier, is 115 km south-west of Bangalore. Located on a river island, this fort has not witnessed urbanization on the same disruptive scale as Bangalore. Further, the fort has three concentric walls with moats adjacent to each wall that are carved deep into the bedrock. Unlike the situation shown in Fig. 2.6a, b, these moats do not get fully silted up and the parts that are not subject to regular conservation are filled with wild vegetation. For these reasons, despite the urbanized landcover, the moats and fort bastions are easily visible in both the synoptic and ground views (Fig. 2.16).

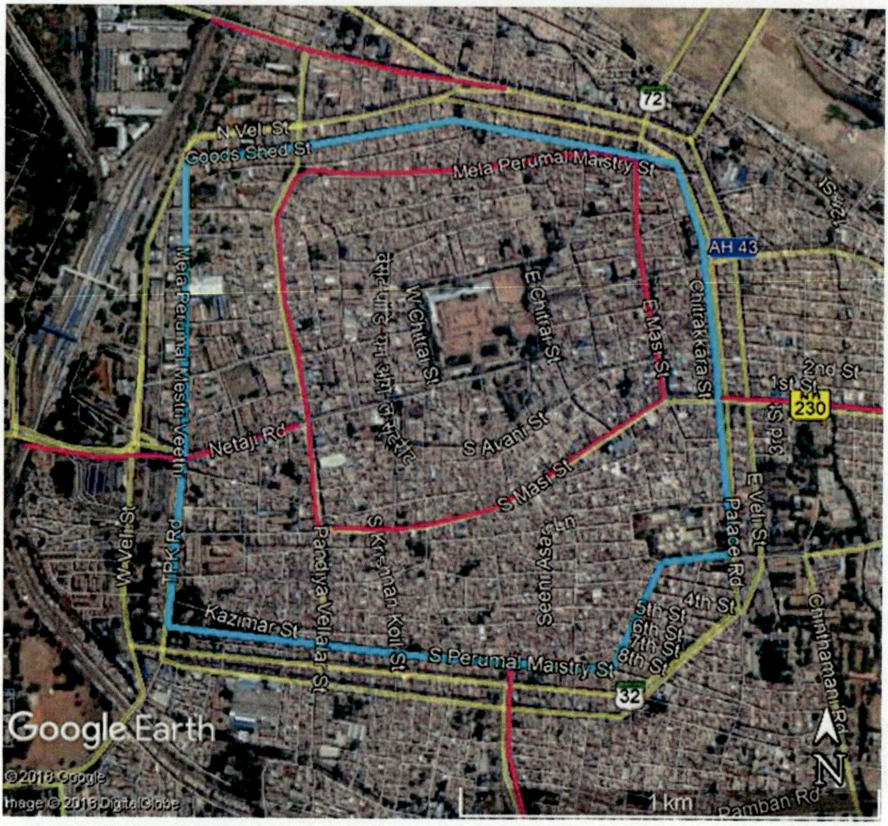

Fig. 2.19 Fort feature from *Plan of Madura* (1755) overlaid on Google Earth (fort walls are marked in cyan and major roads are in magenta)

2.3.4 Settlement Mounds in Rural Settings

The accumulated refuse generated by people living on the same site for several hundred years forms artificial mounds. If the settlement is abandoned, further layers of dust and silt can be deposited by wind or floods on these mounds. These mounds can sometimes assume the shapes of archaeological structures or layouts buried within them, provided they contain enough volume of intact material. Such mounds stand out for their anthropogenic shapes in contrast to their surrounding terrain. Figure 2.22a illustrates an example of a settlement which, over time, morphs into a mound (Fig. 2.22b). A later settlement may develop atop this mound, as shown in Fig. 2.22c. Within a larger mound, there can be smaller mounds that contain individual structures (Fig. 2.9c, d). To visualize such features geospatially, one needs to use digital elevation models (DEM) (discussed in Sect. 3.3).

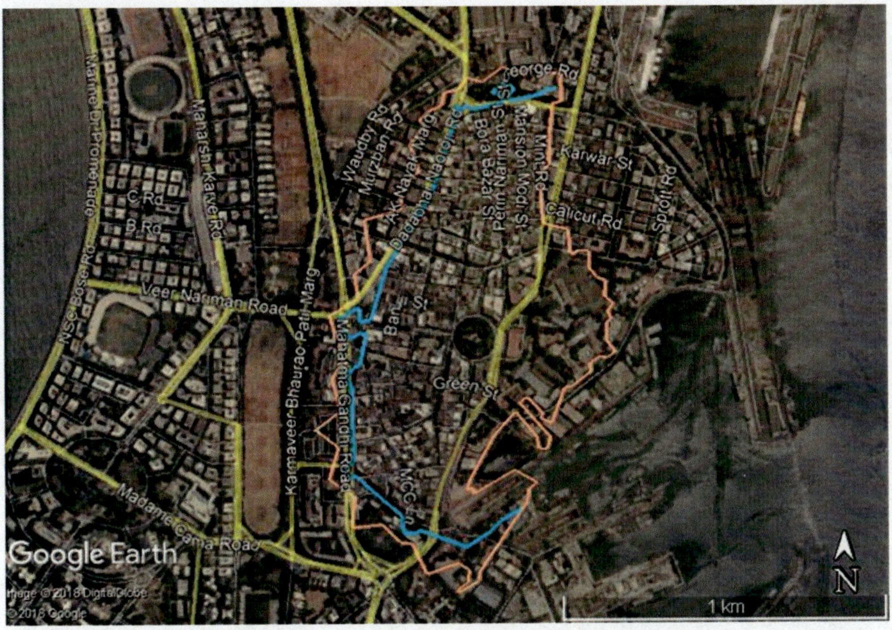

Fig. 2.20 Fort feature from *Bombay Fort 1771–1864* overlaid on Google Earth (fort walls are marked in cyan and the parapet is marked in beige)

2.3.5 Rocky Terrain

Humans have carved out rocks to create sacred structures and dwellings (e.g. the caves of Badami, Ajanta and Ellora). Aerial and satellite images are of little use in studying such structures because there are generally no traces visible from above. However, humans have also built structures on top of rocky terrain (e.g. the upper and lower Sivalaya in Badami). The morphology of rocky terrain is over geological timescales (i.e. much slower than archaeological timescales), and bare rocks do not accumulate silt (it gets washed off by rain). Thus, while such structures may deteriorate, their remains are usually visible on the surface (see Fig. 2.23a, b). Very high spatial resolution images are best suited for detecting such remains, since one is looking for individual structures rather than larger landscape features. When rocks are not bare but have some vegetation, built remains can be obscured and hence harder to identify (see Fig. 2.23c, d).

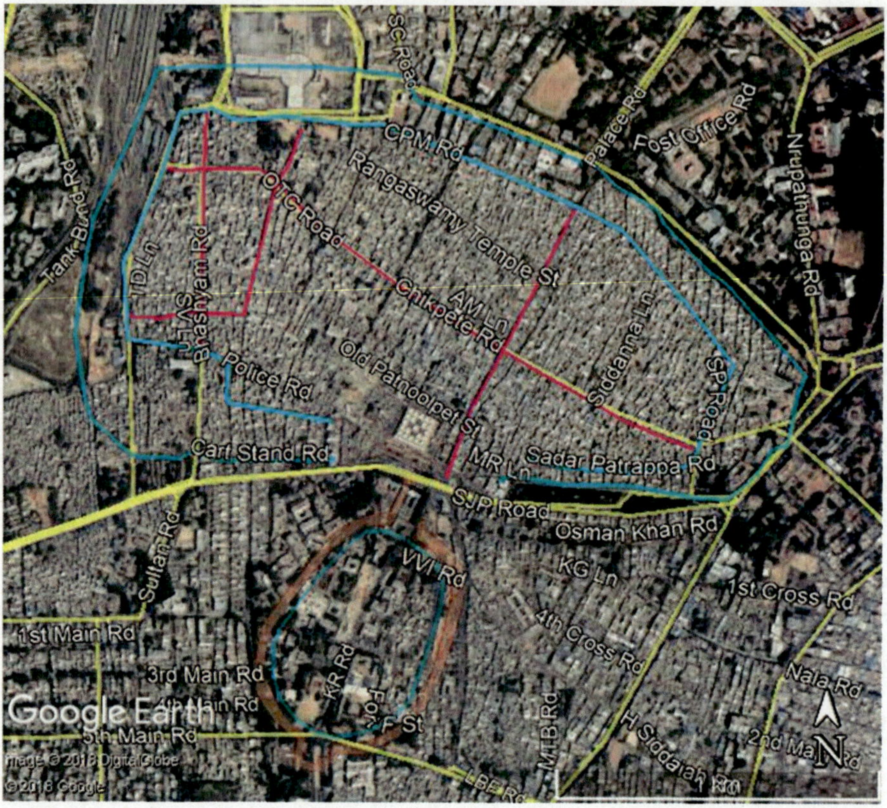

Fig. 2.21 Fort feature from *Plan of Bangalore* (1791) overlaid on Google Earth (fort walls are marked in cyan, major roads in magenta and the moat is marked in beige)

2.3.6 Riverbanks/Floodplains

Riverbanks and floodplains are mostly used for agriculture, which has been discussed in Sects. 2.3.2 and 2.3.4. However, a feature that is often associated with such a setting is a palaeochannel which may have served as a source of water for the settlement. (The source may have dried up or the flow may have changed course subsequently.) Synoptic views provided by satellite images are extremely useful in identifying such channels, which may not be visible while traversing the area on foot. Palaeochannels often have more subsurface moisture than their immediate surroundings, which results in a healthy vegetation similar to the effect that positive cropmarks have (see Sect. 2.3.2 and Fig. 2.7a), but the pattern has a distinct riverine shape and covers a larger area. A palaeochannel can stretch for many kilometres and span several fields and a variety of landcovers.

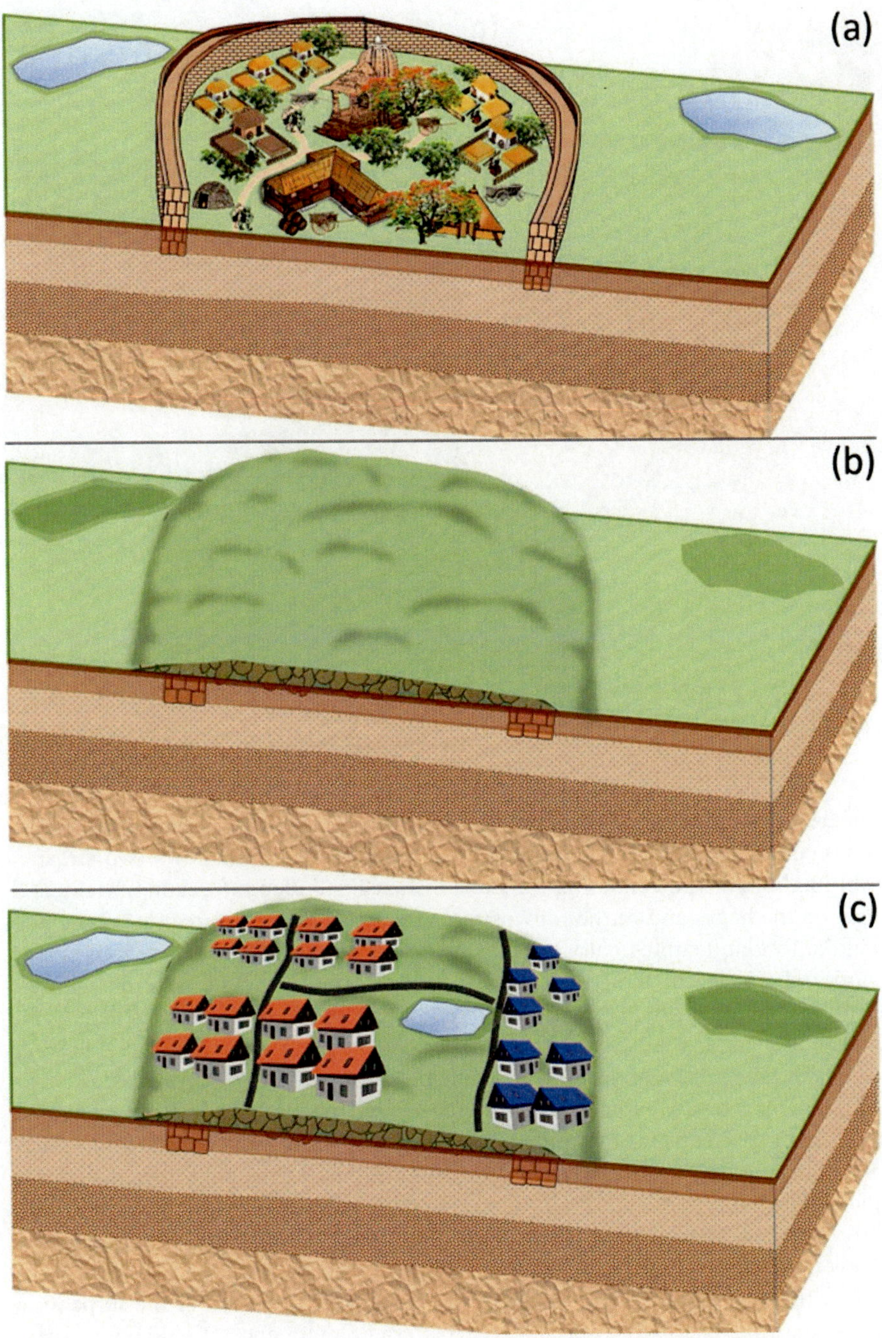

Fig. 2.22 Diagram showing the morphology of a settlement mound

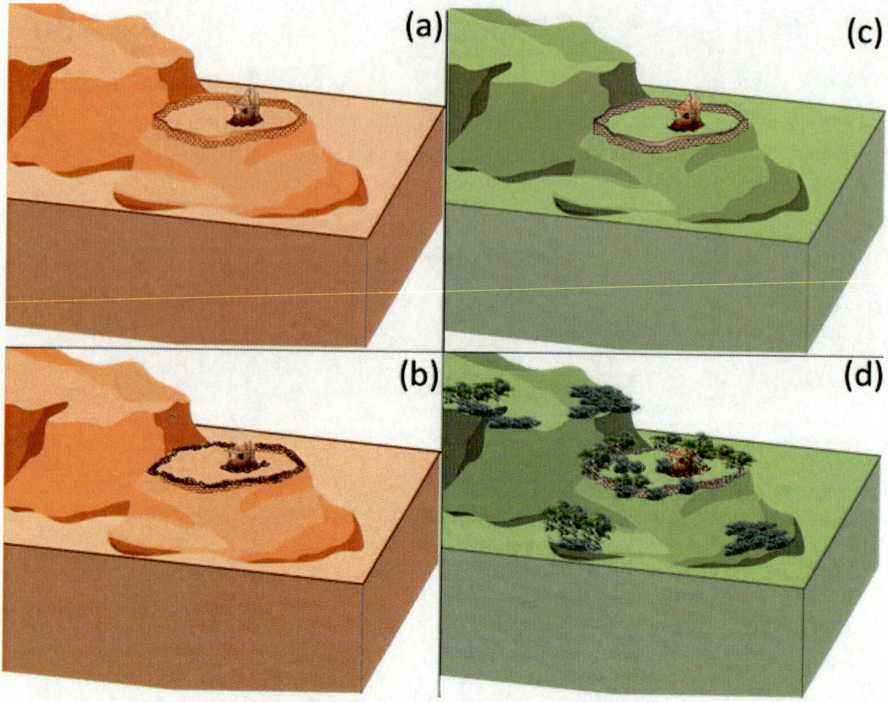

Fig. 2.23 Diagrams showing structures on rocky or hilly terrain

Figure 2.24 illustrates how palaeochannels manifest among agricultural fields: let us imagine a river flowing from right to left. Figure 2.24a shows two streams converging into one, meandering for some distance and then bifurcating. There is a temple on the bank, and further downstream (adjacent to the bifurcation) there is a triangular fort built strategically close to the flow. Figure 2.24b depicts a subsequent scenario where one of the tributaries has dried up and the other one has shifted slightly. The temple, which was originally on the riverbank, now stands isolated, and the fort no longer appears to have been deliberately constructed close to the flow. A good example of such morphology is seen in the crescent-shaped fortified settlement of Sravasti, the site of Buddha's Jetavana (the second monastery donated to Gautama Buddha after the Venuvana in Rajgir). Here, the concave curve on the north-west side would have followed the meandering river, which would earlier have flowed adjacent to the abutting fort (Fig. 2.25a, b). Figure 2.24c shows a scenario where the whole section of the channel is inactive and has subsequently been used for agriculture. However, the shape of the stream can be preserved (most often unintentionally) as property boundaries in the surrounding land. Thus, one can identify the serpentine shape of a former channel in the collective pattern of a sequence of field boundaries. A variation of this is depicted in Fig. 2.24d. Here, the spatial patterning of parts of the palaeochannel is not as distinct as in Fig. 2.24c because the contours of the parcels have obscured the meandering profile. The distinctness of the meandering shape may

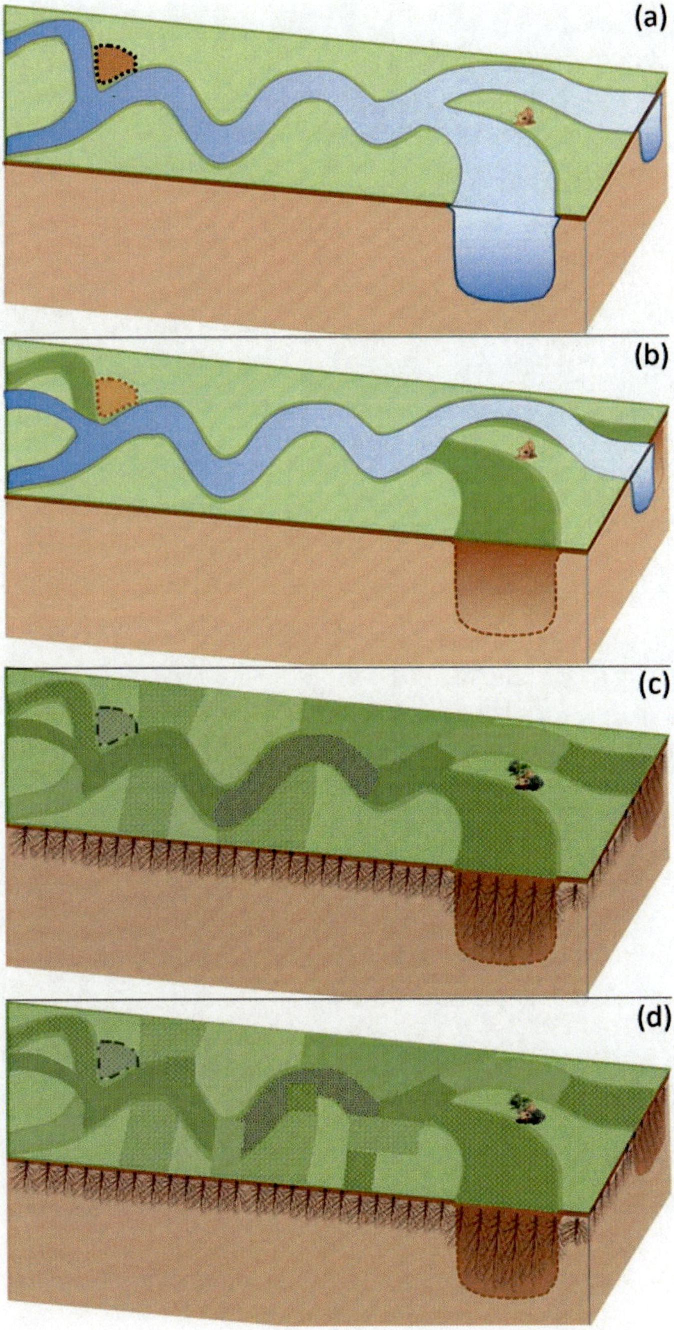

Fig. 2.24 Diagram showing how palaeochannels manifest among agricultural fields over time

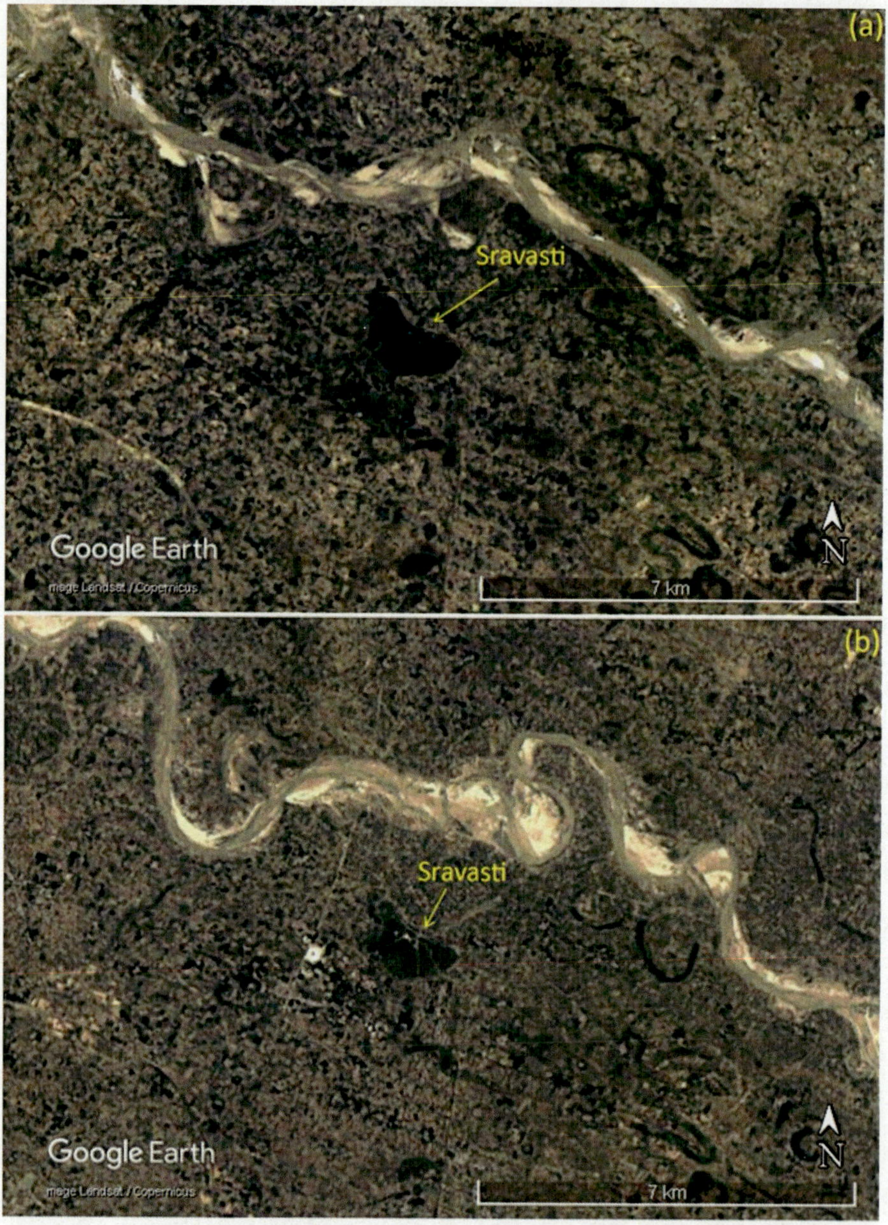

Fig. 2.25 Crescent-shaped fortified site of Sravasti in Uttar Pradesh. The shape of the meandering river as seen on December 1984 in (**a**); and December 2014 in (**b**)

also be subject to the season when the image was taken (i.e. before/after rains and annual crop cycle in the agricultural fields). For instance, a palaeochannel north-east of the site of Sarnath is visible in Fig. 2.26a, but it is less conspicuous in an image taken just two months later (Fig. 2.26b). Figure 2.26c, d show a snaky pattern in a sequence of field boundaries, but we don't see a positive cropmark because most of the parcels along the palaeochannel were fallow on the dates when these images were taken. Hence, it is very important to analyse images from multiple dates and seasons.

Channels can vary in width from a few metres for small streams and rivulets to several kilometres for large river systems, and s correspondingly vary in width. The palaeochannels of a major river system that flowed from the Himalayas to the Kutch have been identified (Rajani and Rajawat 2011). Several Harappan sites are located along the banks of this former river and its tributaries (see Sect. 4.1). Figure 2.3a shows a section in northern Rajasthan overlaid with the locations of Harappan sites.

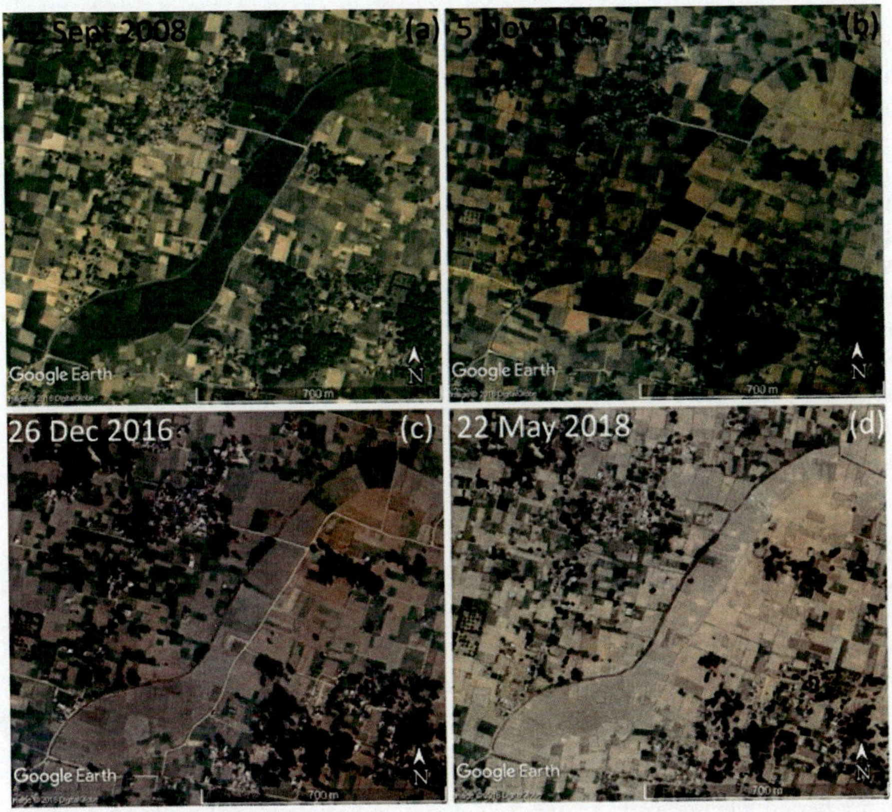

Fig. 2.26 Variations in the visibility of a palaeochannel north-east of Sarnath seen on four different dates

2.3.7 Coastal Regions

Many major cities across the world have been located along the coast, which provided easy access to resources, trade, and mobility via the seas. Historical spatial records often identify coastal structures in relation to the coast. This is unfortunate because the coast is characterized by the constant interaction of terrestrial processes and marine processes such as erosion, deposition, and storm surges. These processes heavily influence the spatial patterns we see in spatial contexts of coastal sites, which makes it difficult to match historical spatial records with present-day coastlines.

To appreciate this difficulty, consider the hypothetical coastal site shown in Fig. 2.27 with six structures, marked 1–6. Let us assume we have records for the time when the site was as depicted in Fig. 2.27a, where structures 5 and 6 were on the shore, structures 2, 3, and 4 were inland from the shore (structure 4 had a surrounding low wall), and structure 1 was still further inland. Figure 2.27b shows a different coastline, caused by some combination of erosion and sea level rise. Historical records from this time may fail to note structures 5 and 6 (because they are completely submerged) and may only note the structures 2, 3 and 4 located on the shore (the latter within an enclosing wall). Figure 2.27c shows further changes to the coastline, and records made at this time would indicate four coastal structures: 2, 3, 4 (partly eroded wall) and 5 (which has resurfaced but is in ruins). The apparent inconsistencies between these three historical reports are clearly due to the dynamic movements of the coastline, so the focus when reviewing these reports must be on immovable features: the built structures themselves, as well as large rocks, roads, etc.

Some of the earliest modern maps of India were coastal maps dating to the colonial period (sixteenth century CE onwards), when Portuguese, Dutch, French, Danish, and English colonists and traders settled on the Indian coasts. Sea charts and maritime maps facilitated the safe transportation of goods and the protection of settlements from rivals and were therefore guarded with great secrecy. Today, these archived maps provide archaeologists a wealth of spatial information for coastal sites.

As an example, consider the site of Mahabalipuram. This port city has many monuments from the Pallava dynasty dating from the seventh to the ninth centuries CE. Seven free-standing temples were visible near the shore to maritime travellers who sailed past this site during medieval times, and this unique landmark gave the site the toponym *Seven Pagodas*. However, from at least 1788 (Carr 1984) to today, only one temple stands close to the shoreline (there are more temples inland, and submerged ruins as well). This has led to much speculation about which seven monuments were being referenced, and whether they are on land or are submerged. A Dutch portolan chart from 1670 may resolve this mystery. It marks seven shrines along a coastline whose shape differs from the modern coastline (Fig. 2.28). Since this map dates from a period when the site acquired its name *Seven Pagodas*, it could be used to identify these seven monuments (see Sect. 5.5) (Rajani and Kasturirangan 2013). However, such maps can be subject to errors, biases, and limitations of technology of the time. These aspects of maps and the challenges involved in using them are further discussed in Sect. 4.3.3.

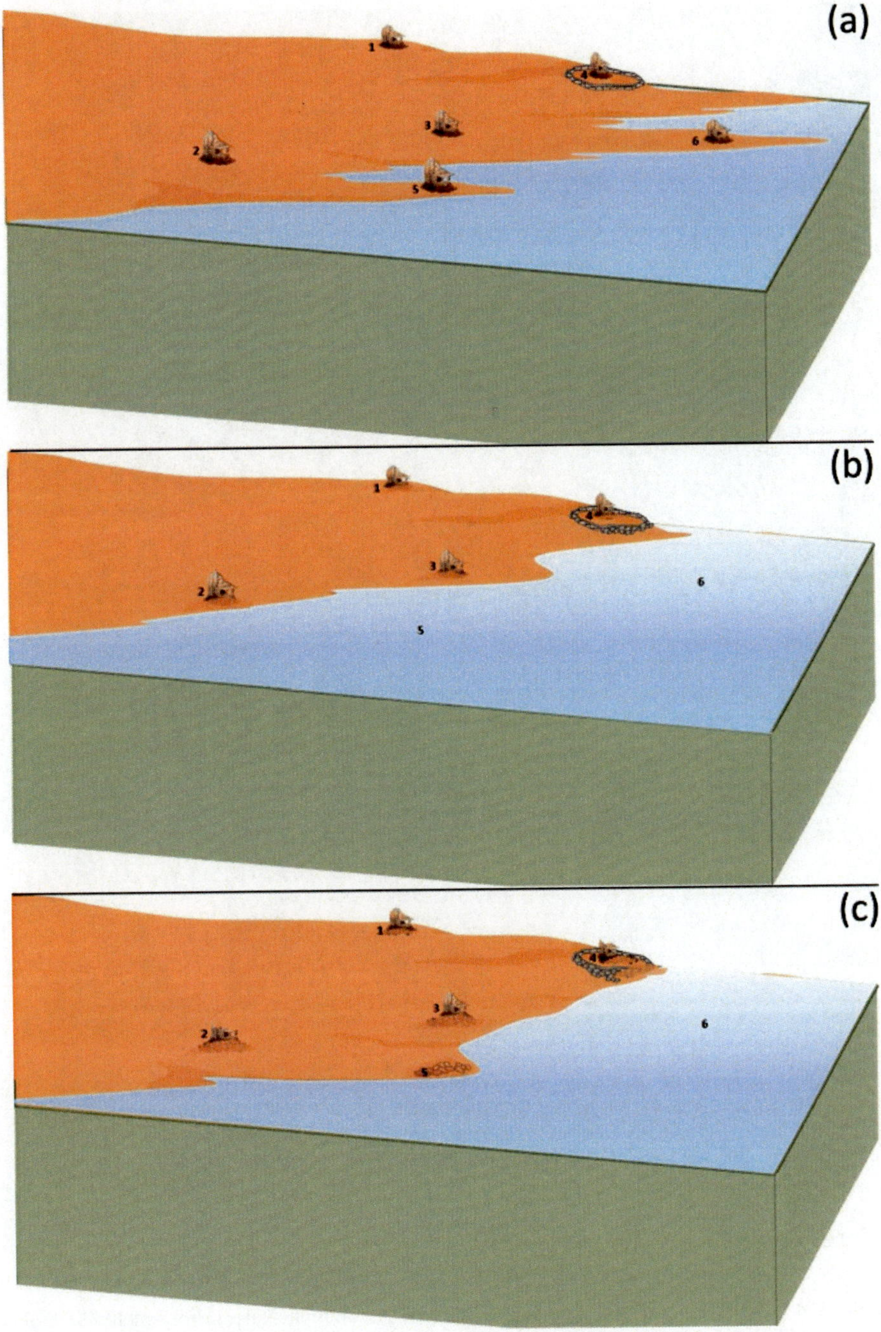

Fig. 2.27 Diagram showing a hypothetical coast with six structures at three different times

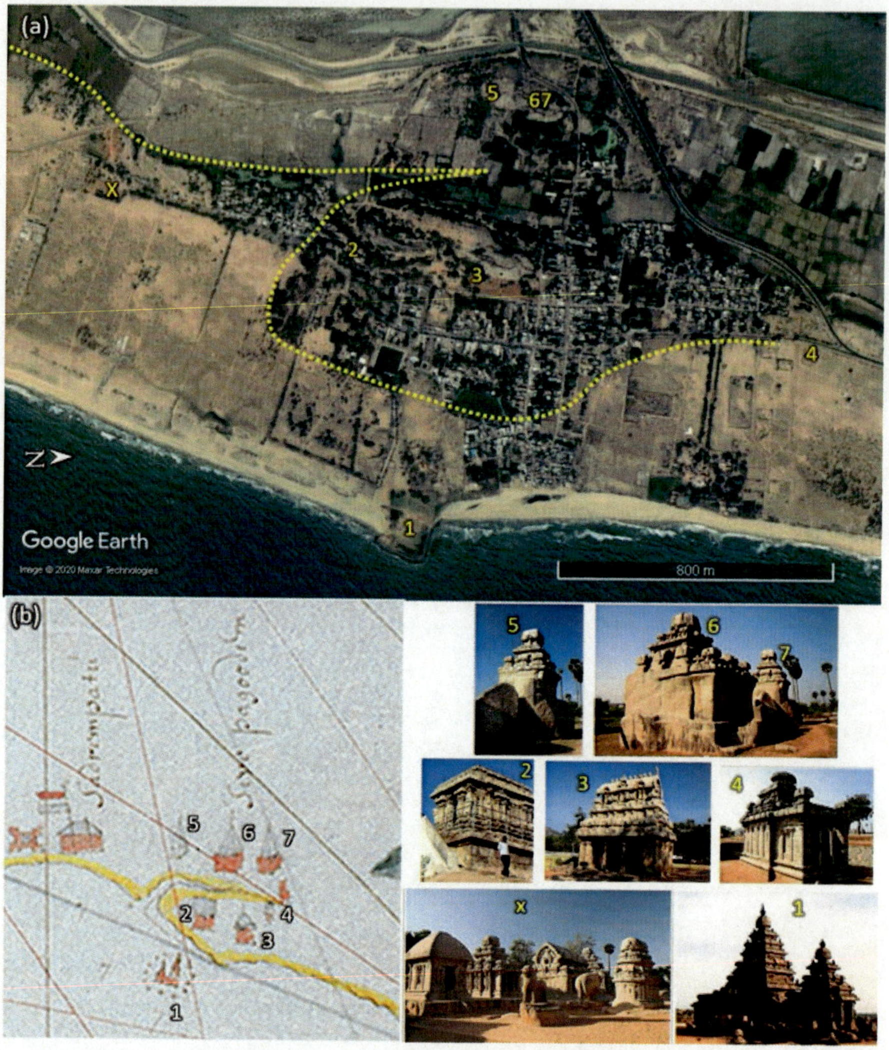

Fig. 2.28 **a** Locations and distribution of free-standing monuments at Mahabalipuram as seen on Google Earth image; **b** a portion of the Dutch Portolan chart of 1670 showing the monuments; 1—shore temple, 2—Olakkanatha or light house temple, 3—Ganesha ratha, 4—Mukunda nayanar temple, 5—Valiyankuttai ratha, 6 and 7—the two Pidari ratha, X—the five Pandava ratha

References

Beveridge H (1900) The fort of Masulipatam in 1759. A comprehensive history of India, civil, military, and social, from the first landing of the English to the suppression of the Sepoy revolt; including an outline of the early history of Hindoostan, London Blackie:614. https://archive.org/stream/comprehensivehis01beve/comprehensivehis01beve#page/614/mode/1up. Accessed 12 Apr 2020

Bhadreenath S, Achyuthan H, Haricharan S, Mohandas KP (2011) Saluvankuppam coastal temple-excavation and application of soil micro morphology. Curr Sci 100(7):1071–1075

Bradford J (1957) Ancient landscapes: studies in field archaeology. G. Bell and Sons Ltd, London

Carr MW (ed) (1984) Seven pagodas on the coromandel coast: descriptive and historical papers By Chambers, Goldingham, Babington, Mohan, Braddock, Taylor Elliot & Cubbins. Asian Educational Services, New Delhi

Das S, Rajani MB (2020) A geospatial study of archaeological remains at Halebidu: an integrative approach to identify unexplored features. Journal of Indian Society of Remote Sensing, submitted, unpublished

Iyer M (2019) Discovering Bengaluru. INTACH Bengaluru Chapter: 44

Rajani MB (2007) Bangalore from above: an archaeological review. Curr Sci 93(10):1352–1353

Rajani MB, Kasturirangan K (2013) Sea-Level changes and its impact on coastal archaeological monuments: seven pagodas of Mahabalipuram, a case study. J Indian Soc Remote Sens 41, 461–468. https://doi.org/10.1007/s12524-012-0210-y

Rajani MB, Kasturirangan K (2014) Multispectral remote sensing data analysis and application for detecting moats around medieval settlements in South India. J Indian Soc Remote Sens 42, 651–657. https://doi.org/10.1007/s12524-013-0346-4

Rajani MB, Rajawat AS (2011) Potential of satellite based sensors for studying distribution of archaeological sites along palaeo channels: Harappan sites a case study. J Archaeolog Sci Elsevier 38(9) Sept:2010–2016. https://doi.org/10.1016/j.jas.2010.08.008

Rouhi J (2016) The seismicity of Iran and its effect on Iranian adobe cultural heritage: the case study of Bam City. In: The 4th international congress on civil engineering, architecture & urban development Tehran, Iran, 27–29 Dec

Soil Horizon. https://www.britannica.com/science/horizon-soil. Accessed 12 Apr 2020

Thakker PS (2001) Locating potential archaeological sites in Gujarat using satellite data. NNRMS Bull 26 (Department of Space, Govt. of India, Bangalore)

Trumpler C (ed) (2005) The past form above. Francis Lincoln Limited, London

Wilson DR (2000) Air photo interpretation for archaeologists. Tempus, Gloucestershire

Chapter 3
The Science and Technology of Remote Sensing in the Context of Archaeology

3.1 Imaging Sensors

A schematic representation of the relation between an analyst who is interested in performing geospatial analysis using remote sensing imagery, the sensors which captures these images, and the process of creating raster images is shown in Fig. 3.1. Clearly, the analyst must have sufficient understanding of the science of imaging, the capabilities and associated terminology of sensors, and the steps involved in image processing to search for and acquire imagery suitable for their analysis. The purpose of this chapter is to help readers without expertise in remote sensing develop this understanding.

Figure 3.2 categorizes sensors based on the information they provide specifically for applications to cultural heritage studies. The top-level categories are images of surface (and subsurface) and topography and 3D models. The former category of sensors is further split into three groups based on the wavelength ranges in which they operate: (1) wide bands in the visible and infrared regions, (2) narrow hyper-spectral bands, and (3) microwave bands. Sensors within these groups offer different combinations of spatial resolution (which places a lower limit on the size of objects one can typically observe with such sensors) and spectral resolutions (the number of spectral bands that a sensor can detect and the width of the regions of the electromagnetic spectrum for each band). Sensors in the latter top-level category (topography and 3D models) are categorized based on their spatial resolutions, the type of data they produce, the kinds of sensors that provide such data, and sources for procuring such data. The number at the end of each arrow in Fig. 3.2 references a particular sensor, whose details are provided in Table 3.1. This table is not meant to be exhaustive—it is merely indicative of the kinds of sensors available today. As sensor technologies improve, the types of sensors and the categorization presented in Fig. 3.2 may change. Further, as the spatial resolution of sensors improves, analysts may be able to observe objects smaller than buildings.

M. B. Rajani, *Patterns in Past Settlements: Geospatial Analysis of Imprints of Cultural Heritage on Landscapes*, Springer Remote Sensing/Photogrammetry,
https://doi.org/10.1007/978-981-15-7466-5_3

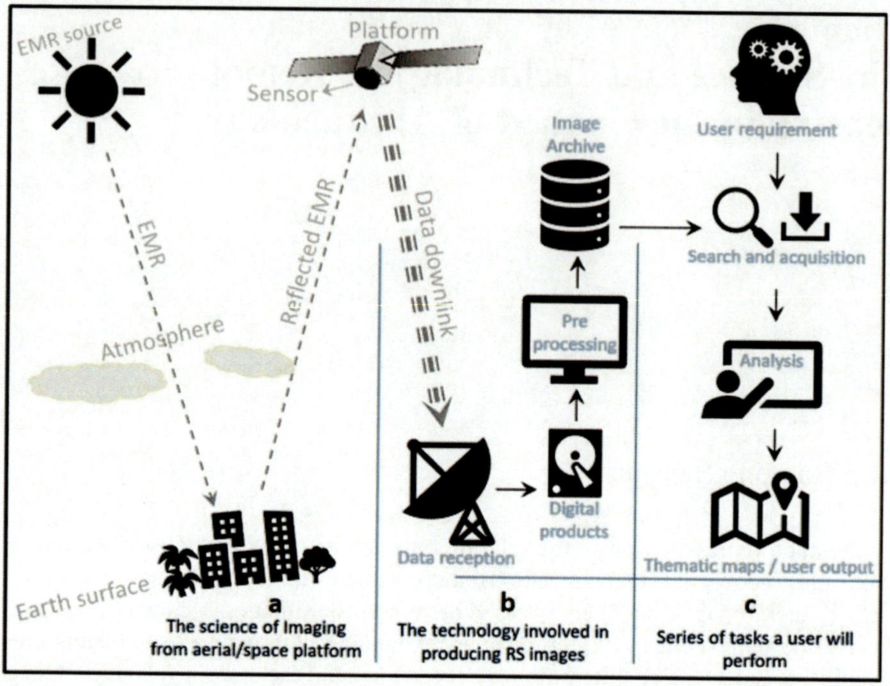

Fig. 3.1 Systemic diagram showing the process of **a** image acquisition; **b** image preprocessing; **c** analysis

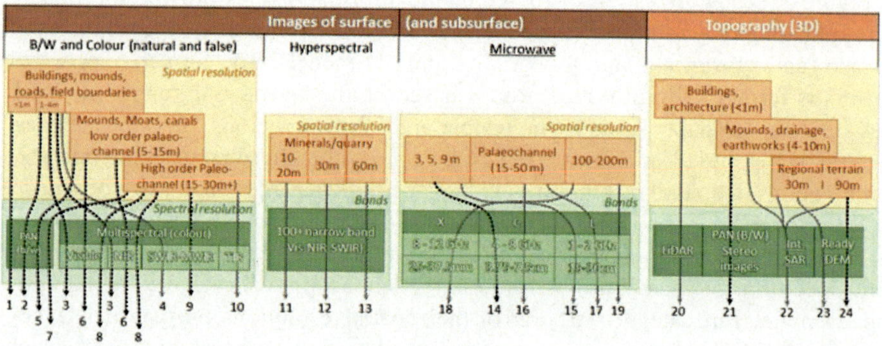

Fig. 3.2 Categories of sensors based on features of interest for cultural heritage studies. The arrows point to numbers indicating sensors listed in Table 3.1

Active and passive sensors:

Passive sensors detect natural radiation, i.e. either reflected solar radiation or radiation emitted by the earth's surface. In contrast, active sensors generate electromagnetic radiation (at specific wavelengths, or across a band of wavelengths) to illuminate

Table 3.1 Some of the Earth Observation Satellites that produce remote sensing imagery (Refer to Fig. 3.2 for the numbers in the first column)

BLACK AND WHITE AND COLOUR (NATURAL AND FALSE) IMAGES				
	Spatial Res. (m)	*Bands*	*Satellite (Launch) Sensor*	*Swath (Km)*
1	Panchromatic (Spatial Resolution <1 m)			
	0.31	PAN	WORLDVIEW-3 (2014)	13.1
	0.31	PAN	WORLDVIEW-4 (2016)	13.1
	0.46	PAN	WORLDVIEW-2 (2009)	16.4
	0.46	PAN	GEOEYE-1 (2008)	15.2
	0.50	PAN	WORLDVIEW-1 (2007)	17.2
	0.61	PAN	QUICKBIRD (2001)	16.8
	0.7	PAN	PLEIADES-1A (2011)	20
	0.7	PAN	PLEIADES-1B (2012)	20
	0.7	PAN	KOMPSAT-3A (2015)	16.8
	0.82	PAN	IKONOS (1999)	11.3
	0.83	PAN	CARTOSAT-2 (2007)	~10
2	Panchromatic (Spatial Resolution 1-4 m)			
	2.5	PAN	CARTOSAT-1 (2005) PAN AFT	30
	2.5	PAN	SPOT 5 (2002) PAN	60
3	Multispectral (Visible and NIR); Spatial Resolution 1-4 m			
	1.24	8: Blue′, Blue, Green, Yellow, Red, Red, NIR^1,NIR^2	WORLDVIEW-3 (2014)	13.1
	1.24	4: Blue, Green, Red and NIR	WORLDVIEW-4 (2016)	13.1
	1.84	8: Blue′, Blue, Green, Yellow, Red, Red, NIR^1,NIR^2	WORLDVIEW-2 (2009)	16.4
	2.44	4: Blue, Green, Red and NIR	QUICKBIRD (2001)	16.8

(continued)

Table 3.1 (continued)

	Spatial Res. (m)	*Bands*	*Satellite (Launch) Sensor*	*Swath (Km)*
	2.8	4: Blue, Green, Red and NIR	PLEIADES-1A (2011)	20
	2.8	4: Blue, Green, Red and NIR	PLEIADES-1B (2012)	20
	2.8	4: Blue, Green, Red and NIR	KOMPSAT-3A (2015)	16.8
	3.2	4: Blue, Green, Red and NIR	IKONOS (1999)	11.3
4	Multispectral sensor (SWIR-MWIR); Spatial Resolution 1-4 m			
	3.70	SWIR (8 bands)	WORLDVIEW-3 (2014)	13.1
	5.5	MWIR 3.3–5.2 μm	KOMPSAT-3A (2015)	16.8
5	Panchromatic (Spatial Resolution 5-10 m)			
	5	PAN	SPOT 5 (2002)	60
	5.8	PAN	IRS-1C (1995)	70
	5.8	PAN	IRS-1D (1998)	70
	10	PAN	SPOT 1 (1986)	117
	10	PAN	SPOT 2 (1990)	117
	10	PAN	SPOT 3 (1993)	117
	10	PAN	SPOT 4 (1998)	60
6	Multispectral images (Visible and NIR); Spatial Resolution 5–10 m			
	5.8	3: Green, Red and NIR	RESOURCESAT-1 (2003) LISS-4	70
	5.8	3: Green, Red and NIR	RESOURCESAT-2 (2011) LISS-4	70
	5.8	3: Green, Red and NIR	RESOURCESAT-2A (2016) LISS-4	70
	6	4: Blue, Green, Red and NIR	SPOT 6 (2012) MS	60
	6	4: Blue, Green, Red and NIR	SPOT 7 (2014) MS	60
	10	4: Blue, Green, Red and NIR	SPOT 5 (2002) MS	60

(continued)

Table 3.1 (continued)

	Spatial Res. (m)	*Bands*	*Satellite (Launch) Sensor*	*Swath (Km)*
7	PAN (Spatial Resolution 15-30+ m)			
	15	PAN	LANDSAT 6,7,8 (1993-99-2013) ETM, ETM, OLI	185
	40	PAN	LANDSAT 3 (1978) RBV	185
8	Multispectral (Visible and NIR); Spatial Resolution 15-30+ m			
	15	3: Green, Red and NIRI	TERRA (1999) ASTER	60
	20	4: Blue, Green, Red and NIR	SPOT 4 (1998)	60
	23.5	4: Blue, Green, Red and NIR	IRS 1C, 1D (1995,97) LISS -3	140
	30	4: Blue, Green, Red and NIR	Landsat 4-8 (1982, 84, 93, 99, 2013) TM, TM, ETM, ETM+, OLI	185
	36.25	4: Blue, Green, Red and NIR	IRS 1A, 1B, P2 (1988,91,93) LISS -2	71
	72.5	4: Blue, Green, Red and NIR	IRS 1A, 1B, P2 (1988,91,93) LISS -1	148
	56	4: Blue, Green, Red and NIR	RESOURCESAT-1,2,2A (2003,11,16) AWiFS	740
	79	4: Blue, Green, Red and NIR	Landsat 1-3 (1972, 75, 78) RBV, MSS	185
	188	4: Blue, Green, Red and NIR	IRS 1C, 1D (1995,97) WiFS	770
9	Multispectral sensor (SWIR-MWIR); Spatial Resolution 15-30+ m			
	20	SWIR (1.58 to 1.75 mm)	SPOT 4 (1998)	60
	23.5	SWIR	IRS 1C, 1D (1995,97) LISS -3	140
	23.5	SWIR	RESOURCESAT-1,2,2A (2003,11,16) LISS -3	140
	30	SWIR x2 (1.57 - 1.65 μm) and (2.11 - 2.29 μm)	LANDSAT 8 (2013) OLI	185
	30	MWIR (2.08 - 2.35 μm)	LANDSAT4-5 (1982, 84) TM, TM	
	30	MWIR (2.08 - 2.35 μm)	LANDSAT 7 (1999) ETM+	185

(continued)

Table 3.1 (continued)

	Spatial Res. (m)	*Bands*	*Satellite (Launch) Sensor*	*Swath (Km)*
	30	SWIR x 6	TERRA (1999) ASTER	60
	56	SWIR	RESOURCESAT-1,2,2A (2003,11,16) AWiFS	740
10	Multispectral sensor (TIR); Spatial Resolution very coarse			
	60	TIR (10.40 - 12.50 μm)	LANDSAT 6-7 (1993-99) ETM, ETM+	185
	79	TIR (10.4 to 12.6 μm)	LANDSAT-3 (1978) MSS	185
	90	TIR x 5	TERRA (1999) ASTER	60
	100	TIRS 1 (10.6 - 11.19 μm) TIRS 2 (11.5 - 12.51 μm)	LANDSAT 8 (2013) TIRS x 2	185
	120	TIR (10.40 - 12.50 μm)	LANDSAT4-5 (1982, 84) TM, TM	185
HYPERSPECTRAL				
11	10-2	19: 400-1050 nm	PROBA-1 (2001) CHRIS	14
12	30	220: 357-2576 nm	Earth Observing-1 (2000) Hyperion	7.7
13	60	13: 443–2190 nm	Sentinel-2 (2015) MSI	290km
MICROWAVE				
	Nominal Resolution (m)	*Frequency (GHz)*	*Satellites*	*Swath (km)*
14	3 x 3 to 5 x 50	C (5.3)	RISAT-1 (2012)	10-225
	5 x 5 to 25 x 100	C (5.405)	SENTINEL-1A & 1 B (2014, 16)	80 to 400
	9 x 9 to 100 x 100	C (5.3)	RADARSAT – 1 & 2 (1995, 2002)	45/510
	10-200	C 5.289	SIR-C (1994 & 95)	15-90
15	3	L (1.257)	ALOS- 1 & 2 (2006, 14)	25
16	26 x 28	C (5.3)	ERS-1/2 (1991)	100
	30	C (5.3)	ENVISAT-ASAR (2002)	56

(continued)

Table 3.1 (continued)

	Nominal Resolution (m)	*Frequency (GHz)*	*Satellites*	*Swath (km)*
17	25	L (1.275)	SEASAT (1978)	100
	40	L (1.275)	SIR-A (1981)	50
	15-50	L (1.275)	SIR-B (1984)	20-50
	18 x 18	L (1.275)	JERS-1 (1992)	75
18	10-200	X 9.602	SIR-C (1994 & 95)	15-90
19	10-200	L 1.239	SIR-C (1994 & 95)	15-90
TOPOGRAPHY FROM DEM				
20	LiDAR (<1m spatial resolution)			
21	Stereo images			
	Spatial res. of DEM (m)	*Satellite*	*Sensors/tilt*	*Swath (km)*
	4-10	CARTOSAT1 B/H Ratio: 0.64	Fore +26 AST -5	30
	4-10	ALOS B/H Ratio: 1	PRISM Nadir (N) Forward (f/w): +24 Backward(b/w): -24	N: 70 f/w:35 b/w:35
22	Interferometric SAR			
	Spatial res. of DEM (m)	*Satellites*	*Band*	
	12, 30, 90	TerraSAR-X (TSX)	X	
	10	Sentinel-1	C	
	10-30	ALOS-2	L	

(continued)

Table 3.1 (continued)

	Ready DEM		
	Spatial res. of DEM (m)	*Satellites*	*Source*
23	30	CARTOSAT1	https://bhuvanapp3.nrsc.gov.in/data/download/
	30	ASTER	https://earthdata.nasa.gov/learn/articles/new-aster-gdem
	30	SRTM	https://earthexplorer.usgs.gov/
24	90	SRTM	http://srtm.csi.cgiar.org/

surface objects and then detect the scattered radiation reflected from them. As an illustration, the use of a flash-bulb can turn a passive photographic camera sensor into an active sensor. Sensors that observe in optical and infrared regions of the electromagnetic spectrum are passive sensors because they image reflected sunlight. In contrast, sensors that observe microwave or LiDAR are active sensors. For clarity, active sensors are underlined in Fig. 3.2.

3.1.1 Sensor Parameters

Sensors are further characterized by their spatial resolution, the number of spectral bands in which they operate, their bandwidth, their radiometric resolution, and their repetivity/revisit (the duration between two consecutive observations of the same location by the sensor) and their swath (the extent or width of the earth's surface, in kilometres, observable by a sensor at one time). Recall that observations from space platforms provide synoptic views that allow large swathes of area to be viewed simultaneously.

Spatial resolution: Spatial resolution is the minimum separation between two objects at which they can be distinguished on an image. A sensor with high spatial resolution can image (record) so that closely spaced objects appear as separate objects. A digital image is a matrix of pixels (picture elements), and each pixel corresponds to a square area on the ground (see Sect. 4.2). In an image produced by a sensor with 1 m spatial resolution, each pixel corresponds to a 1 m • 1 m square on the ground. Such an image shows finer details of buildings, roads, and other features than an image produced by a sensor with coarser resolutions (e.g. 5.8 m, 23.5 m, or 30 m per pixel). The latter show features that are part of the larger landscape, such as drainage patterns and old water bodies. As we noted in Sect. 2.2, such features can (somewhat counterintuitively) be harder to perceive at higher resolutions.

Spectral resolution: It is difficult for a sensor to detect reflected/emitted radiation across a continuous part of the electromagnetic (EM) spectrum. Instead, sensors are designed to sense reflected radiation within a few specific wavelength ranges or bandwidths (carefully chosen to be in the atmospheric window, yet away from the absorption bands of atmospheric constituents), and these bandwidths define the sensor's spectral resolution. The visible part of the EM spectrum (i.e. visible light) consists of radiation in the narrow range of approximately 0.4 μm (micrometer) to approximately 0.7 μm. This can be split into the Blue spectrum (roughly 0.4–0.5 μm), the Green spectrum (0.5 μm to 0.6 μm), and the Red spectrum (0.6–0.7 μm). (We capitalize Blue, Green, and Red to distinguish these spectra from the colours blue, green, and red.) Due to differences in their physical properties, the reflectance in the Blue, Green, and Red parts of the spectrum differs for each object, and hence, they appear to be of different colours (Fig. 3.3). There are similar variations between objects in the invisible parts of the EM spectrum. These variations are captured in the so-called spectral signature or spectral curve of objects (Fig. 3.4). The spectral curve for vegetation is particularly interesting. The reflectance is about 7% in the

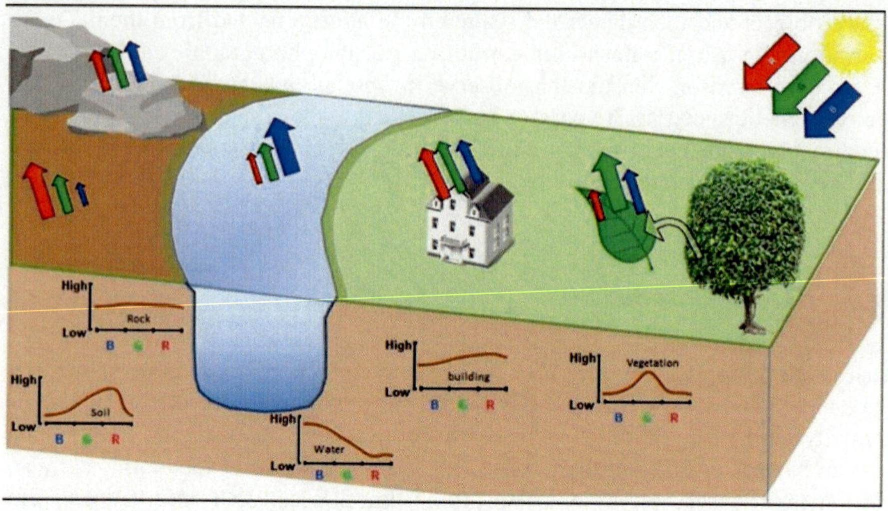

Fig. 3.3 Diagram showing reflectance in the Blue, Green, and Red regions of spectrum from typical earth surface features

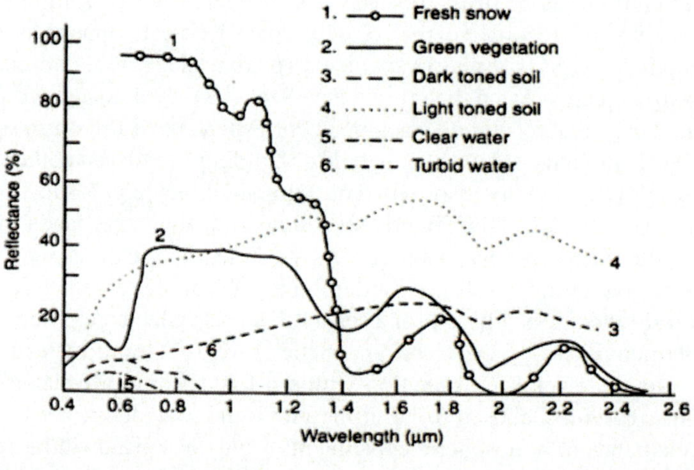

Fig. 3.4 Reflectance curves for different earth surface features (reproduced from Joseph and Navalgund 1991)

Blue band, but this rises to about 14% in Green band and then falls down to 10% in Red band and beyond 0.7 μm (the so-called near infrared or NIR band) it shoots up to more than 40% (due to the internal structure of leaves, as explained in Box 2), stays high before falling down around 1.2 μm. As a result, sensors that can detect in the NIR band are particularly useful for detecting cropmarks (see Fig. 2.8).

Box 1: Electromagnetic (EM) energy is the basis of remote sensing observations. Visible light is only one of the many forms of EM energy. All this energy is inherently similar and radiates in accordance with basic wave theory. EM energy travels in a harmonic, sinusoidal fashion at the "speed of light" ($c = 3 \bullet 10^8$ m/s). The distance from one wave peak to the next is called the wavelength λ (measured in metres or fractions such as μm, mm, etc.), and the number of peaks passing a fixed point in space in one second is called the frequency ν (measured in Hertz or multiples such as kHz, MHz, etc.). Hence, $c = \lambda\nu$ and since c is a constant, it follows that λ are ν inversely related. This can be seen visually in Fig. Box 1a.

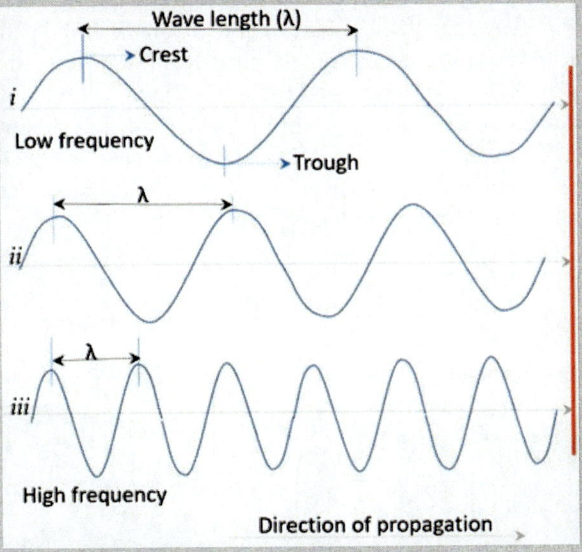

Fig. Box 1a Relation between wavelength and frequency of electromagnetic radiation

Waves *i*, *ii* and *iii* are moving at the same speed *c*. If two crests/troughs of wave *i* strike the red line in a unit of time (say 1 s), then three and six crests/troughs of waves *ii* and *iii* would, respectively, strike in the same duration. Among the three waves, wave *i* has the lowest frequency and highest wavelength, whereas wave *iii* has the highest frequency and lowest wavelength.

Electromagnetic radiation covers a wide range of frequency (or wavelengths—either term can be used to characterize a wave), extending from gamma rays (with wavelengths around 10^{-10} m) to radio waves (with wavelengths more than 1 m). Although there are no clear divisions between any two regions of the EM spectrum (Fig. Box 1b),

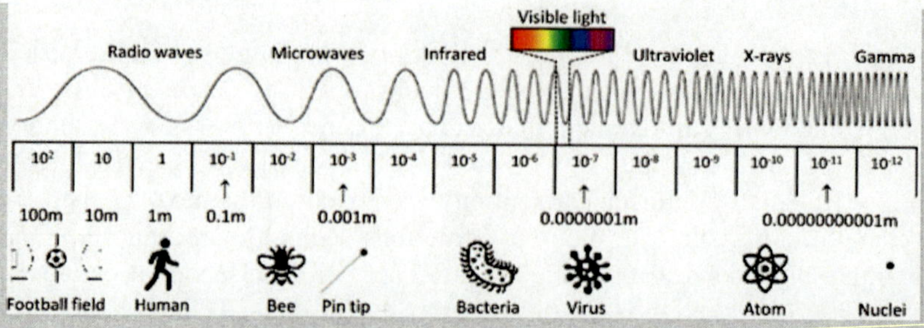

Fig. Box 1b Electromagnetic spectrum

it is useful to differentiate between regions according to wavelengths: the ultraviolet region (less than 0.4 μm), the visible region (0.4–0.7 μm), the three categories of infrared (IR) regions namely near IR (0.7–1.3 μm), mid IR (1.3–3 μm) and thermal IR (3–14 μm), and the microwave region (1 mm to 1 m). Intermittent ranges of EM radiation are blocked by the earth's atmosphere, so earth observation from space is not possible in these ranges. The range of wavelengths in which the atmosphere is transparent are called atmospheric windows (Fig. Box 1c). They strongly influence the choice of wavelength while designing a sensor system.

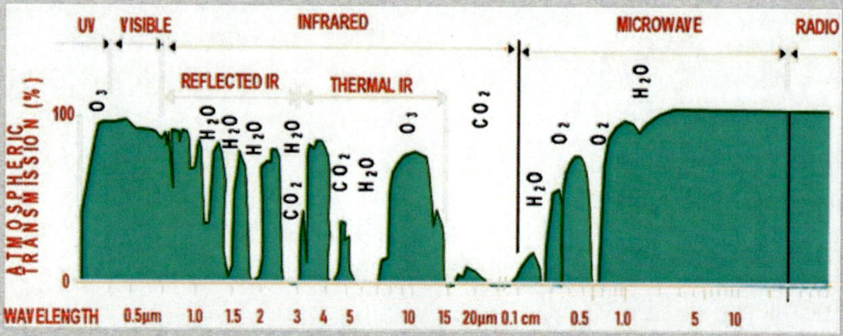

Fig. Box 1c Atmospheric window (adapted from Sabins 1997). Electromagnetic radiation is blocked in the regions where the atmospheric transmission (Y-axis) is low. Hence, remote sensing from space platforms is infeasible in these regions of the spectrum

The interaction of EM radiation with matter is also characterized by particle model, which models waves as discrete packets of energy (or quanta) called photons. The frequency of the wave is proportional to the magnitude of the particle's energy. Moreover, because photons are emitted and absorbed by charged particles, they act as transporters of energy. The energy per photon can be calculated by Planck's equation: $E = h\nu$, where E is the energy in Joules (J), $h = 6.626 \bullet 10^{-34}$ Js is Planck's constant, and ν is the frequency (Hz). Combining this equation from the particle model with the equation $c =$

$\lambda\nu$ from the wave model, we obtain $E = hc/\lambda$. Hence, short-wavelength (high frequency) waves have higher energy than long-wavelength (low frequency) waves.

In addition, objects also emit radiation as per Planck's law (Joseph and Jeganathan 2018), and again the energy involved is inversely proportional to the wavelength of the radiation. This theory has important implications for remote sensing: the longer the wavelength naturally emitted by terrain, the lesser the energy and hence the more difficult it is for a sensor to gain a detectable signal. To detect such radiation, the sensor would have to observe a large area to gather sufficient low-energy radiation, leading to coarser spatial resolutions or will have to observe an area for longer duration.

Box 2: The internal structure of leaves is responsible for high NIR reflectance from vegetation. The structure of a typical leaf consists of two primary layers: the palisade mesophyll (tightly packed elongated cells) above and spongy mesophyll (irregular shaped cells with air gaps in between) underneath, sandwiched by thin layers of cells (epidermis) on either side (see Fig. Box 2).

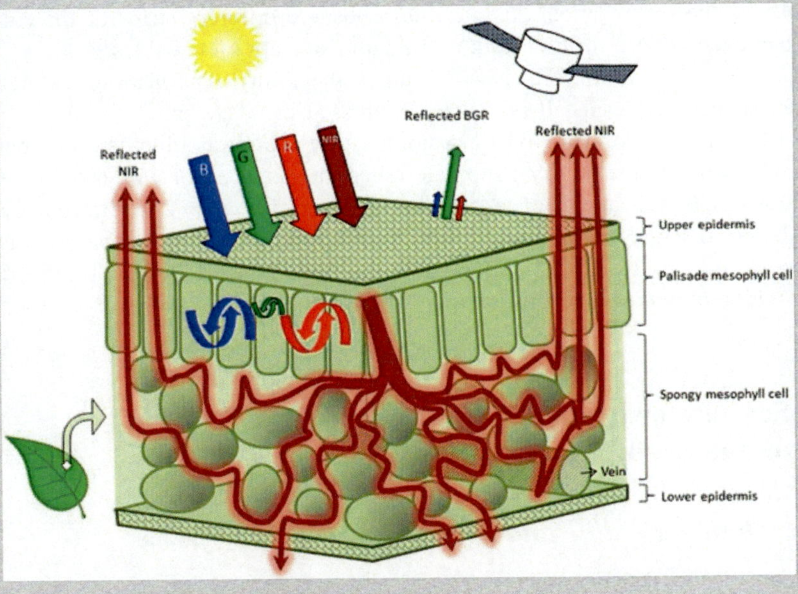

Fig. Box 2 Diagram indicating pathways of light through a leaf

Leaves appear green to us because of all the light in the visible region that falls upon them, a large portion of energy in the Green region is reflected by pigments in the top layer. The energy in the Blue and Red regions is largely absorbed by the palisade layer to stimulate photosynthesis. Energy from other regions of spectrum behaves differently while interacting with various parts of a leaf's structure. NIR energy penetrates the top layer, undergoes refraction in the lower spongy layer as it interfaces with cells of varied sizes and shapes and the interim space, and it gets dispersed. This behaviour of NIR makes it more diffuse than the light in the visible region, which only interacts with the top layer (Joseph and Jeganathan 2018). The NIR energy that is reflected upwards is then imaged by infrared sensors on satellites. Healthy plants have more leaves of larger size, creating a larger leaf surface area to reflect NIR energy than for unhealthy plants. Therefore, the distinction between healthy and less healthy vegetation is sharper in an NIR image (https://science.nasa.gov/ems/08_nearinfraredwaves).

Temporal resolution: One of the main advantages of satellite remote sensing is the ability of a satellite to repeatedly observe a specific location at near-regular time intervals. This temporal frequency of a satellite (usually expressed in days) is called its temporal resolution. The satellite orbit is fixed in inertial space making it orbit nearly in the same orbit plane, but since the earth is rotating under this orbit, it makes possible for the sensor aboard a satellite to observe different strips of the earth's surface in each orbit, returning back to the same area after a few days.

Radiometric resolution: This is a measure of the ability of a sensor to distinguish between reflectance levels. It is usually expressed in bits, typically ranging from 8 to 14 bits. Each additional bit of radiometric resolution doubles the number of levels into which the sensor can separate reflection, so a sensor with a radiometric resolution of 8 bits is capable of separating reflectance into $2^8 = 256$ different levels. (In a black-and-white image, these levels are represented as different shades of grey, ranging from black to white.) The number of levels jumps to $2^{14} = 16{,}384$ for a sensor with a radiometric resolution of 14 bits.

3.2 Sensors for Recognizing Archaeological Patterns on the Earth's Surface

3.2.1 Optical (Visible and Infrared)

Electro-optical sensors/camera systems aboard satellites such as LANDSAT, SPOT and IRS have produced images in the visible spectrum (about 0.4–0.7 μm), near infrared (about 0.75–1.4 μm), and shortwave infrared (about 1.4–3 μm) bands by

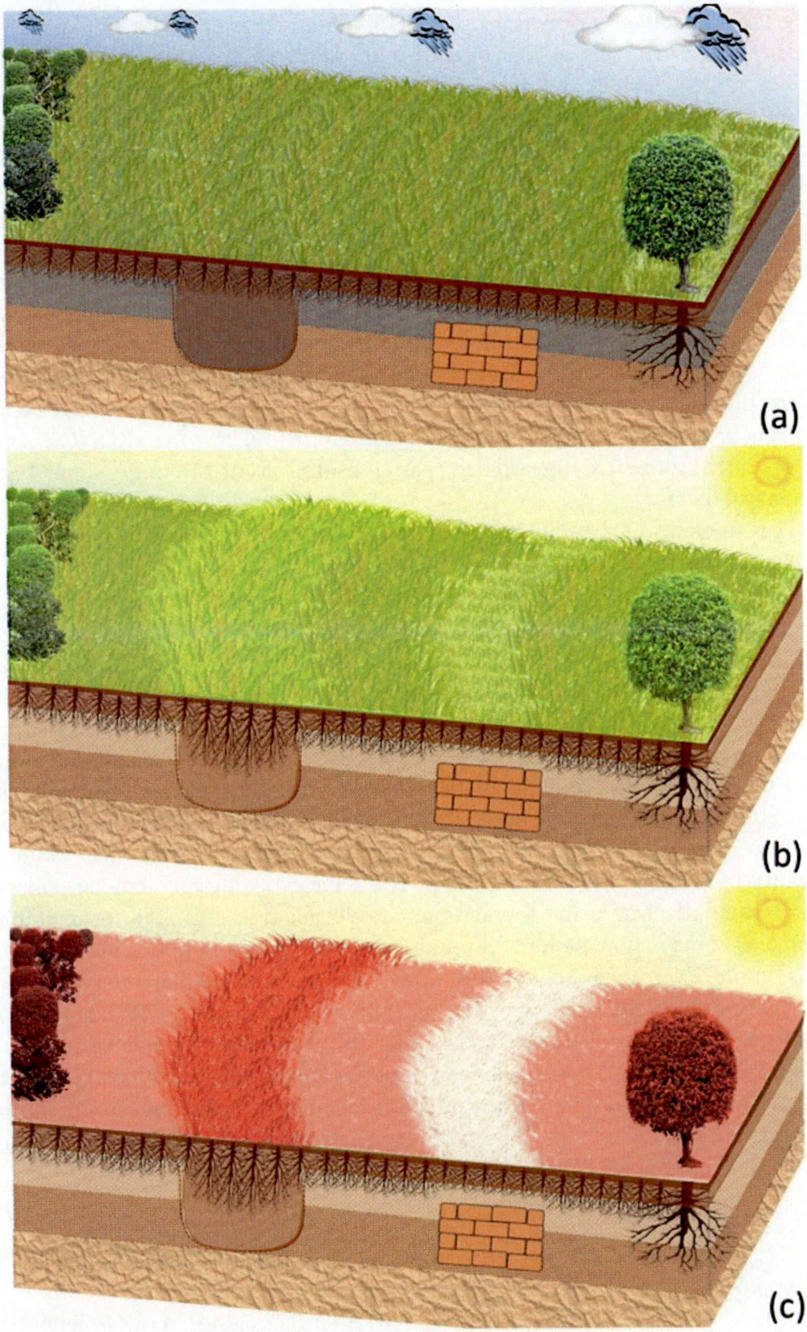

Fig. 3.5 Diagram showing seasonal variability of cropmarks **a** versus **b**; **c** enhanced visibility using infrared image

sensing the respective wavelengths. LANDSAT also obtained data in the thermal infrared (about 3–14 μm) region. Images with high spatial resolution are typically available only for visible and near-infrared bands. For longer wavelengths (which have lower energy, as discussed in Box 1), the field of view must be expanded to sense enough energy to convert into pixel information. A panchromatic (PAN) image is one greyscale image produced from observing the extended visible region (0.3–0.9 μm). A multispectral image consists of multiple greyscale images produced by observing narrower bands (e.g. 0.4–0.5 μm, 0.5–0.6 μm, and 0.6–0.7 μm) and a combination of these bands is used to produce a composite colour image. Typically, at any given phase of sensor technology development, PAN sensors have higher spatial resolution than multispectral sensors because they sense across a larger bandwidth. For instance, IRS 1C and 1D satellites (launched in the 1990s) had PAN sensors with spatial resolutions of 5.8 m and multispectral sensors (green, red and NIR bands) with a much coarser spatial resolution of 23.5 m. Similarly, the IKONOS series which operated from 2000 onward had spatial resolutions of 1 m in PAN and 4 m in multispectral bands.

ASTER operates in multiple bands of the thermal infrared region. Hyperspectral sensors operating in very fine spectral bands (5–10 nm bandwidth) are also flown on some spacecraft missions, which provide spectroscopic information of targets, but in much coarser spatial resolutions.

Buried archaeological features often affect the health of surface vegetation, creating positive or negative cropmarks, which reveal themselves as large patterns when viewed synoptically (see Sect. 2.3.2). The identification of cropmarks is one of the most common techniques for detecting buried archaeological remains (Bradford 1957). Such a signature, together with its shape, size, pattern, texture, and association with adjacent features, helps in identifying and discriminating the object. However, in certain weather conditions, such as post rainfall when there is ample availability of water in the topsoil, it is possible the crops do not show any marks. This is particularly true if the foundations are deeper (see Fig. 3.5a for example). The same area may show patterns through subtle variations in hotter months when water is scarce. Even so, these variations may be so understated that it may be necessary to analyse false-colour composite (FCC) images that include NIR bands before patterns are identifiable. The larger percentage of reflection from vegetation in such images creates a sharper contrast between healthier and normal growth plants (Navalgund and Rajani 2017). Figures 3.5c and 2.8c are simulated examples of the sharper contrast visible in a false-colour infrared image as compared to a visible band image for the same area (Figs. 3.5b and 2.8b).

Box 3: Multispectral sensors are designed to image in specific bands of the EM spectrum. The image for a single band can be rendered in black and white, where regions of high reflection are closer to white and regions of low reflection are closer to black. Figure Box 3

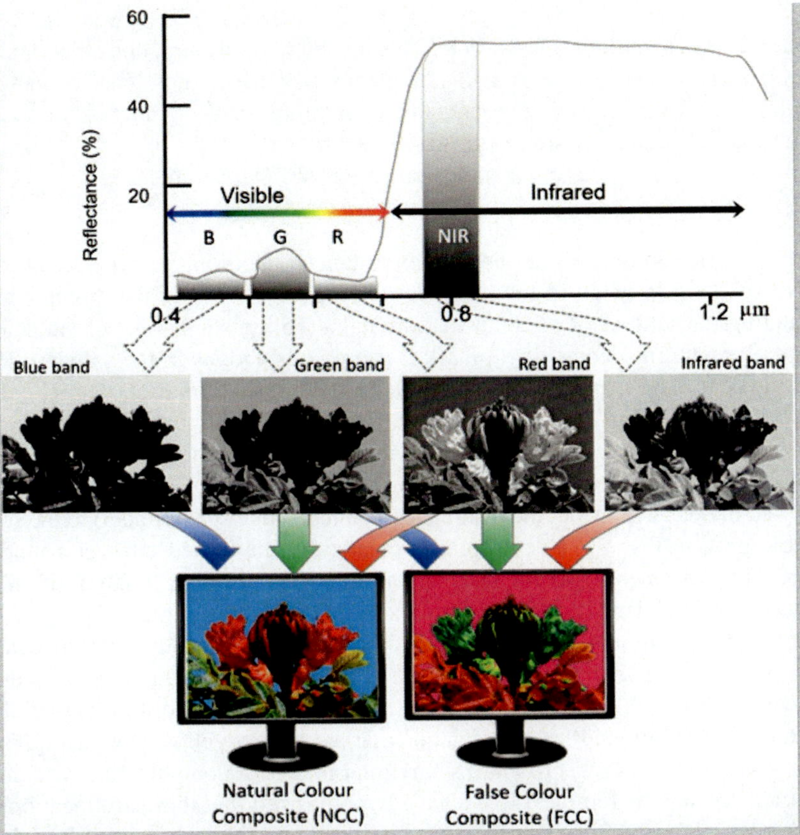

Fig. Box 3 Natural and false-colour composites

uses an example of a picture of a red flower and green leaves against blue sky to illustrate this. The graph shows four bars depicting Blue (B), Green (G), Red (R), and near-infrared (NIR) bands and the corresponding images under them. The Blue band image has high reflectance in the blue region, so the sky looks bright and the rest appears dark. The Red band image has high reflectance from the flower, which appears brighter than the rest of the image. Similarly, leaves appear brighter in the Green band. NIR also produces a black-and-white image showing brighter shades in the region of high NIR reflectance and darker shades in the regions of low reflectance. GIS software enables us to see multispectral images by combining images of three selected bands and assigning primary colours (red, green and blue) to each of the selected bands. If the Blue band image is projected in blue, the Green band image in green, and the Red band image is projected in red, we see a natural colour composite (NCC) image as illustrated in Fig. Box 3.

If we want to include an NIR band image, then we must omit one other band because a colour composite can only be made using three bands. Therefore, we visualize NIR in red, Red in green and Green in blue, making it a false colour composite. Because the reflection from vegetation is high in the NIR band (mapped to red), vegetation appears red colour in FCC satellite imagery.

These subtle differences can be further enhanced using image processing techniques explained in detail in the next section. Figure 3.6 shows the circular moat surrounding the settlement of Belur in Karnataka (15 km south-west of Halebidu). The circular pattern seen as a cropmark is more conspicuous in the standard FCC where the NIR band is projected in red than in the higher-resolution natural colour image from Google Earth (Rajani and Kasturirangan 2014). Figure 3.7 shows the old Bangalore area in a FCC image, which can also be compared against the NCC image in Fig. 2.21. The shape of the *pete* (settlement) area is preserved by road networks (discussed in Sect. 2.3.3), but the oval-shaped pendant-like *kote* (citadel) to the south is visible only in Fig. 3.7 as a cropmark created by the buried moat surrounding the fort. The settlements and environs of Baragaon and Surajpur to the north of the excavated site of Nalanda (seen in natural colour Google Earth images in Fig. 2.12) can similarly be compared against the standard FCC image in Fig. 3.8. The centres of Temples 3, 12, 13, and 14 all lie on an axis line. (Only Temple 14 is seen in Fig. 2.12 and Fig. 3.8.) When this axis line is extended further north, it passes close to an unexcavated mound near Baragaon (marked with a yellow star in Fig. 3.8). Figure 3.8a also shows two cropmarks at regular intervals along this line, which are comparable in size to Temple 14. We have hypothesized that these are locations of past temples whose building materials (surface and possibly subsurface) have been significantly mined to create such cropmarks (Rajani 2016). Note that the cropmarks are more distinct in Fig. 3.8a (taken during a dry period in May 2008) than in Fig. 3.8b (taken in February 2009 when vegetation is widespread due to ample availability of moisture). In the case of such subtle patterns, our confidence in hypothesizing buried structures at these locations stems not just from the cropmarks alone, but also from their clear alignment and regular spacing with known archaeological features.

3.2.1.1 Optical Image Enhancement Techniques

Digital image processing involves manipulating images with the aid of computational image enhancement tools, with the objective of aiding the analyst in interpreting these images by enhancing features of interest. As mentioned earlier, a digital image is a matrix of pixels (picture elements). In a digital/raster image,[1] each pixel has a number (called a Digital Number or DN value, or just DN). In an optical image, this number represents brightness (the minimum value is 0, which indicates darkness). In a digital

[1]Raster is a pattern of closely spaced rows of pixels that form an image.

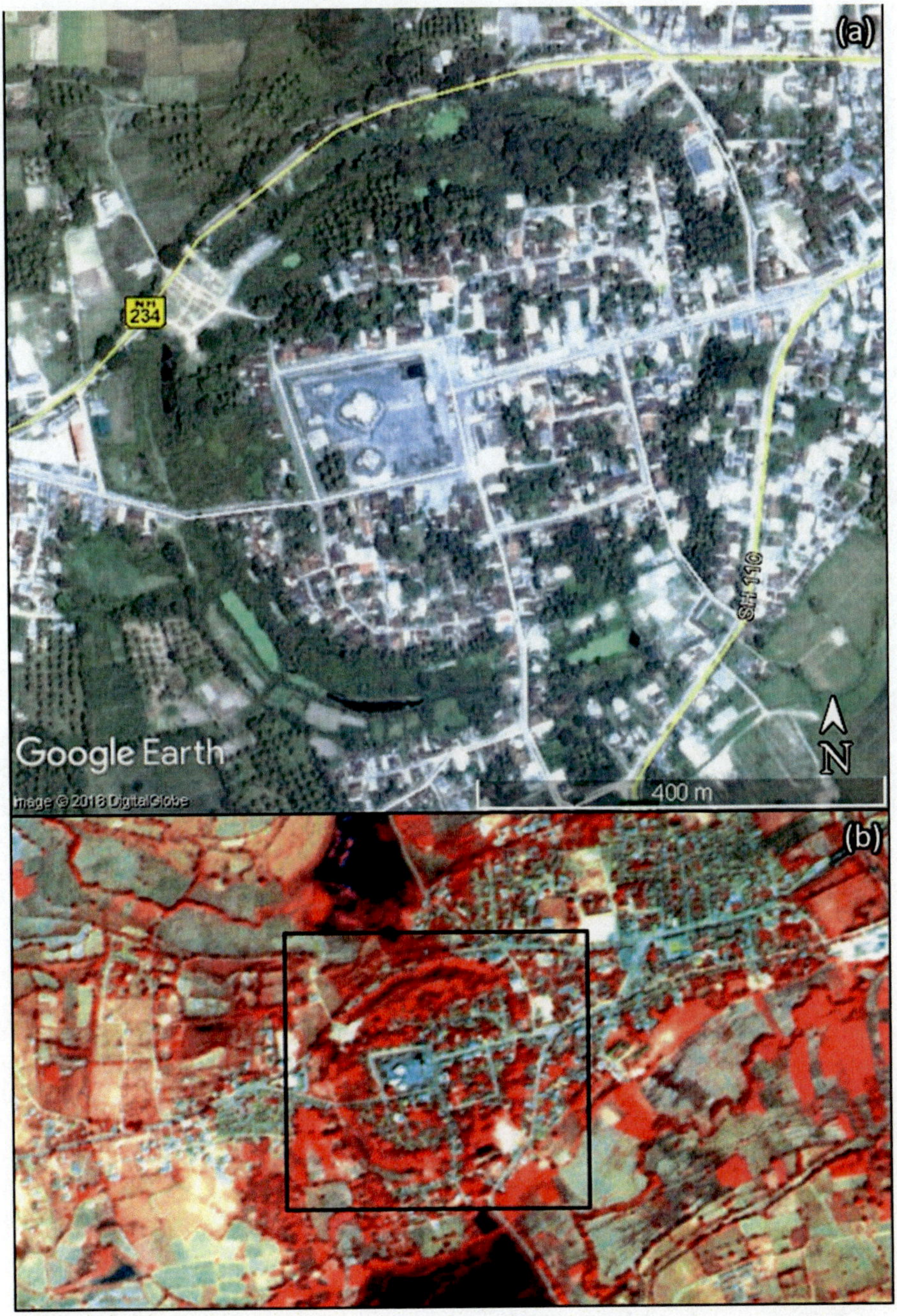

Fig. 3.6 **a** Google Earth image showing the Chennakeshava temple complex in the centre of Belur; **b** the circular moat is more visible in the multispectral image from IRS P6 LISS-4

Fig. 3.7 Old Bangalore area in a multispectral image from IRS P6 LISS-4

elevation model (DEM) (described in Sect. 3.3), this number represents the height of the terrain (typically in metres from mean sea level). Many digital image processing techniques involve substituting the DN associated with a pixel with another DN to enhance a particular feature. To optimize this substitution or to ascertain whether it is even necessary, it is important to first compute the distribution of DNs in the original image. This distribution can be represented as a table showing the frequency of each DN's occurrence. More usefully, this table can be represented visually as a frequency histogram (also known as a DN histogram) as shown in Fig. 3.9a, where the *X*-axis represents the DN and the *Y*-axis represents the frequency of that DN in the image (Gibson and Power 2000). The digital image histogram is also a useful concept for visualizing some of the corrections or enhancements that are applied to digital data (for instance, see histograms in Fig. 3.9b, c).

A priori, it is difficult to know whether any single image enhancement procedure will be suitable. Hence, image-processing software makes it easy for the analyst to experiment with several enhancement procedures in an interactive manner. During this interactive process, the cumulative effects of enhancements to the image

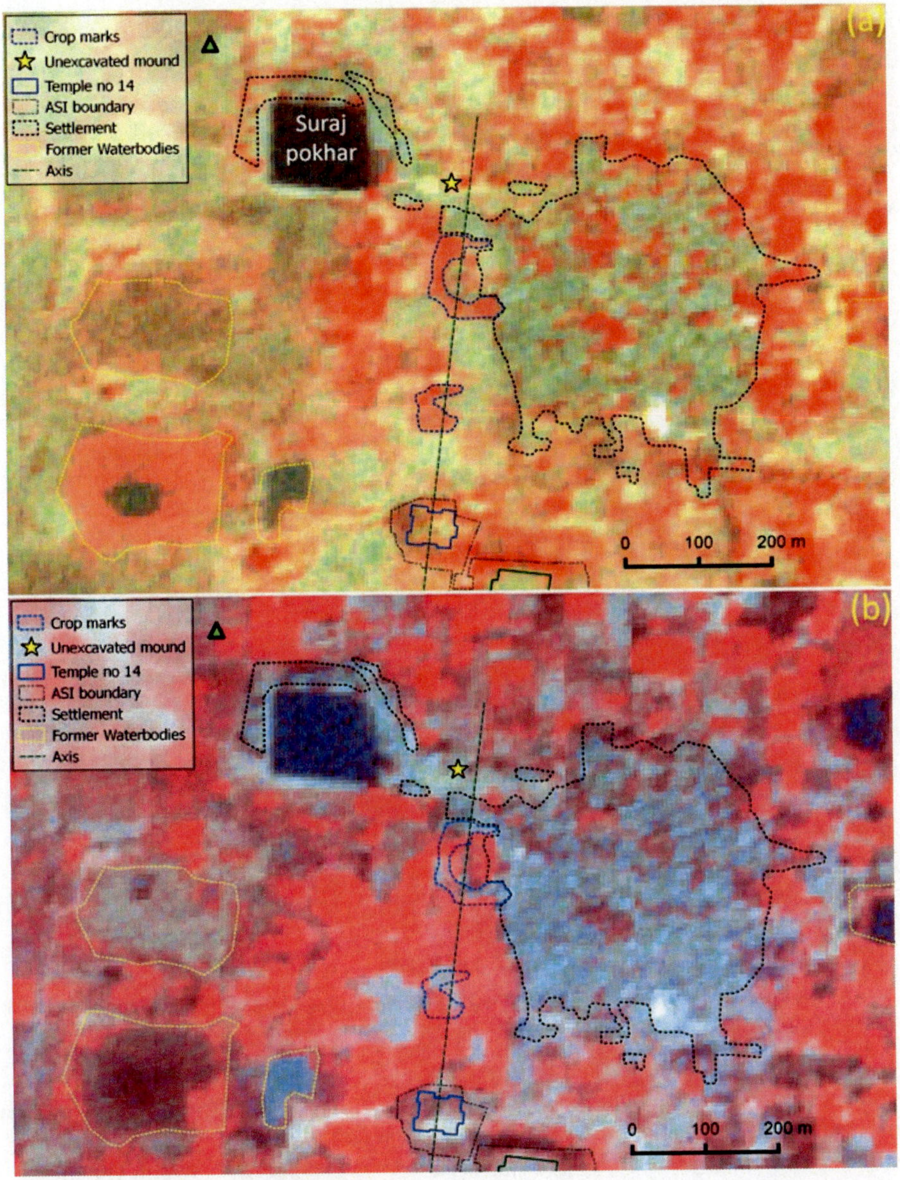

Fig. 3.8 Settlement of Baragaon and Surajpur together with their environs on the north of the site of Nalanda, IRSP6 LISS-4 on **a** 29 May 2008; **b** 12 February 2009

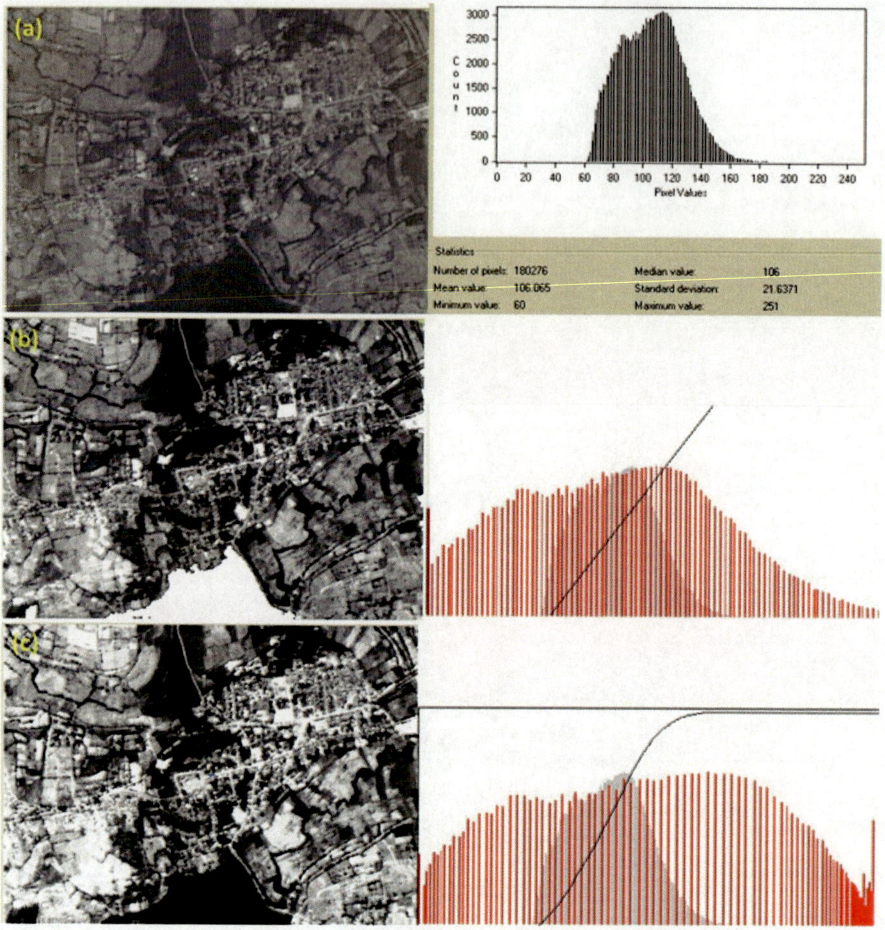

Fig. 3.9 **a** Original single (Green) band image of Belur (left) and its histogram of DN values (right). **b** Image after linear stretch and the corresponding histogram. **c** Image and histogram after nonlinear stretch

are displayed. In the following section, we will confine ourselves to a few image processing techniques that are useful in our domain of interest.

Stretching: Contrast stretching is a process that changes the range of pixel intensity values and can be applied to images with poor contrast. In such images, many natural features within the landscape have a low range of reflectance in a specific wavelength band (and hence, a low range of DN values). To demonstrate this image-processing technique, Band 3 (Green) of IRS P6 LISS-4 of Belur has been used in Fig. 3.9 where it can be difficult to discern features of interest (see Fig. 3.9a). To improve the contrast, the DN values can be "stretched" across a wider range. For instance, if the DN values for the image are in the range 60–175 (as in Fig. 3.9a), these values can be stretched across a wider range (e.g. 0–255). A simple linear stretch can be applied to obtain Fig. 3.9b: the new DN value is obtained by subtracting 60 from the original DN value and then multiplying the result by 255/115 and rounding the result to the nearest integer. It is also possible to perform nonlinear stretches. For instance, a histogram equalization stretch stretches the input data in proportion to the population of the DN bars. This provides a better contrast over the most populated part of the scene (see Fig. 3.9c).

NDVI: Another useful technique is based on calculating the normalized difference vegetation index (NDVI) which can be used to identify subtle cropmarks indicating archaeological features (Belli and Koch 2007; Parcak 2009). The NDVI yields a simple graphical indicator that can be used to analyse multispectral remote sensing measurements and assess whether the target being observed contains live green vegetation or not. For instance, an analysis of NDVI images from Srirangapatna has revealed a linear feature. Subsequent ground truthing (on 6 April 2016) confirmed the presence of two memorial stones marking the end of an esplanade leading to the fort (Fig. 3.10) (Gupta et al. 2017). However, satellite images taken more recently (Fig. 3.10d) shows that clearance for a major road has erased this linear feature highlighting the importance of analysing historical imagery (discussed in detail in Sect. 3.2.3) in the light of rapidly changing landcover.

Classification: The overall objective of image classification procedures is to automatically categorize all pixels in an image into landcover classes or themes. A pixel is characterized by its spectral signature, which is determined by the relative reflectance in different wavelength bands. Multispectral classification is thus an information extraction process that analyses these spectral signatures and assigns the pixels to classes based on similar signatures. After completing such a classification, all pixels belonging to the same class can be assigned a unique colour, resulting in a thematic map.

There are two major approaches to multispectral classification: unsupervised and supervised. These are based on supervised and unsupervised machine learning algorithms. Supervised classification requires the image analyst to provide training samples labelled by class. Thus, the analyst must have the necessary expertise to assign labels correctly. Thereafter, the analyst can choose among the available supervised classification algorithms, and the interested reader can find additional details in Joseph and Jeganathan (2018).

Unsupervised classification, on the other hand, does not require any expertise, or indeed any input (other than the number of classes) from the analyst. The classification technique automatically clusters pixels into the desired number of classes based on the distribution of DN values in the image. Once this is done, however, the analyst must assign meaning to each of the classes. For instance, if the purpose of unsupervised classification is to partition the image into landcover classes, the analyst must assign a class label to each of the unlabelled classes proposed by the classification algorithm.

Fig. 3.10 Two linear features south-east of Srirangapatna fort shown between pairs of yellow and cyan arrows appearing as negative cropmarks in **a** an NDVI image; **b** a false-colour image. These linear features are barely visible in the 2015 Google Earth **c** and not at all in **d** due to clearance for road construction; **e** a ground photograph taken on 6 April 2016 of one of two memorial stones whose locations are marked with yellow stars in (**c**) and (**d**)

GEOBIA: More recently, Geographic Object-Based Image Analysis (GEOBIA) has been proposed as an improvement over pixel-based classification. These techniques recognize that remote sensing images represent distinct objects (with reasonably well-defined boundaries), and hence, the first task is to automatically partition the image into smaller, "meaningful" objects (where the definition of "meaningful" can depend on the domain). Thereafter, these objects can be classified using supervised or unsupervised techniques.[2] While GEOBIA has demonstrated value in domains as varied as military and defence, agriculture, managing natural resources

[2] https://clarklabs.org/segmentation-and-segment-based-classification/. Accessed 12 Apr 2020.

(forest, marine, etc.) and urban studies, its use in archaeology is still quite new (Guio et al. 2015).

3.2.2 Microwave

While the sensors operating in the visible, NIR and thermal regions provide information about surface features, microwave sensors (particularly those operating in the C and L bands—about 5 cm and 21 cm wavelengths, respectively) such as the Synthetic Aperture Radar (SAR) provide subsurface details. SAR is an active sensor, which sends a microwave pulse to the earth's surface and receives a back scattered signal. This response is influenced by the surface object's dielectric properties, its surface roughness, its shape, structure, and orientation. In addition, the response is also influenced by sensor parameters such as its operating frequency, its polarization and the incidence angle. In the microwave region, most natural materials have their dielectric constant between 3 and 8 in dry conditions. Water has a very high dielectric constant of 80. Hence, a change in moisture content causes a significant change in the overall dielectric constant of the soil, resulting in a change in microwave backscatter. The penetration of microwaves is also higher in barren and dry surfaces, and with longer wavelengths. This permits the detection of subsurface (buried) features and palaeochannels. Dykes buried as much as 2 m beneath alluvium (Lavers 2019), unknown fluvial landscapes beneath the Aeolian cover, and the existence of limestone beneath the Aeolian cover has been identified in microwave images (Rajawat et al. 2003). In cases where vegetation cover is present, microwave backscatter depends on the density and geometric structure of the vegetation. At smaller wavelengths (i.e. in the X-band), back scatter is primarily from vegetation. At longer wavelength (i.e. in the L band), it is primarily from the surface. (At intermediate frequencies, it is a mix of both, depending on the angle of incidence.) (Navalgund and Rajani 2017)

One of the early uses of radar for geoarchaeological applications was in detecting drainage patterns under the thick layer of sand in the Sahara (McCauley et al. 1982). Since the dielectric constant in the medium of dry sand is less than soil, the radar pulse can penetrate deeper and reveal the topography of the underlying layer. SAR is also used in interferometric mode to obtain topographical information, as discussed in the next section.

Palaeochannels often have more subsurface moisture than their immediate surroundings, which results in a distinct signature (a channel with a red tinge) in a FCC image. (In some cases, anomalous vegetation growth can produce a similar signature.) Palaeochannels buried under dry sand for a long time may not be readily visible in FCCs. However, microwave SAR (particularly SAR operating at longer wavelengths like L-band) can show such palaeochannels clearly.

In the context of the Harappan civilization, Rajani and Rajawat (2011) have clearly showed that satellite data in both NIR and microwave bands not only helps in identifying palaeochannels, but also in understanding their relationship with settlements. The study showed a large spread of Mature Harappan sites (2200–1700 BC) along the palaeochannel of the Sarasvati and its tributaries in North-West India. In contrast, late Harappan sites (1700–1500 BC) are located further west, in adjoining regions of Pakistan, indicating that the migration of settlements followed the river as it shifted westwards. SAR imagery also revealed a palaeochannel adjacent to the site Talakadu near Mysore (Rajani et al. 2011). Field observations showed agricultural land along this channel, which gets flooded during heavy rains (Fig. 3.11).

Fig. 3.11 Buried palaeochannel seen in **a** RADARSAT SAR Imagery of 22 April 2008; **b** more clearly illustrated on a map; along with photographs from a field survey on 16 August 2008

3.2.3 Historical Imagery

Aerial and satellite images were once highly classified, but today some of these images are easily available. Perhaps the most useful resource of this nature are images taken from Corona satellites, which were used by the US Air Force for photographic surveillance between June 1959 and May 1972. These black-and-white images have worldwide spatial coverage, and they are available for download or can be ordered through USGS Earth Explorer in digital form (https://earthexplorer.usgs.gov/). Corona images have become an important asset in the exploration of hitherto unknown archaeological sites (Jesse 2014; Kennedy 1998; Galiatsatos 2004) for two primary reasons. First, the images are of high resolution (about 1.83–2.74 m), which permits the detection of small features such as former roads, fort-walls, moats, etc. Second, a substantial amount of land was relatively untouched by large-scale mechanized development when these images were taken. Thus, these images contain information that may no longer be visible in satellite imagery.

As an example, Bihar Sharif, the headquarters of Nalanda district (10 km north-west of Nalanda site), had remains of a fort that was surrounded by a wide moat. Bihar Sharif is often associated with Odantapuri (the Buddhist monastery established by Gopala, the first emperor of the Pala dynasty that ruled parts of modern-day northern and north-eastern India, Nepal, and Bangladesh between the eighth and eighteenth centuries CE). The location of Odantapuri has not yet been conclusively established (Rajani and Kumar 2020). Francis Buchanan visited this town on 6 January 1812, and described these massive ruins in his daily journal. He noted that the circular moat was about 600 feet wide on the east and about 400 feet wide on the west (Jackson 1922). Parts of this moat were still visible until the middle of the twentieth century (Patil 1963). Subsequent urbanization has demolished all traces on the ground. However, Corona images taken in 1965 show cropmarks of the wide circular moat, whereas an image of the same region from 2018 shows almost no traces of this circular feature amid the construction of modern buildings (Fig. 3.12). Similarly, Fig. 3.13a (Corona satellite image from October 1965) and Fig. 3.13b (Google Earth image from October 2016) show stark differences in constructed landcover in and around the fortifications of Lalkot and Qila Rai Pithora, in the Mehrauli area of Delhi.

A wealth of valuable historical images may be available from other sources (e.g. the Landsat series of satellites) or may soon become available after it is declassified. In addition, aerial images of several sites are scattered in various collections. For instance, Fig. 3.14 is of Purana Qila in Delhi, aerial photograph taken on 15 February 1946, from Bradford Papers, Pitt Rivers Museum, Oxford, whereas similar photographs of Tughlaqabad in Delhi taken in 1945 and other of Sisupalgarh, Odisha taken in 1948 are also available (Prabhakar and Korisettar 2017). At present, however, the repository of Corona images is the most systematic and easily accessible collection of historical synoptic imagery that we are aware of.

Fig. 3.12 Bihar Sharif **a** Corona image of 1965; **b** the same in GE image of 2018

3.3 Sensors that Facilitate 3D Visualization of Landscapes

Most 3D digital images are in the form of a DEM, also known as a Digital Terrain Model (DTM). A DEM is a digital representation of ground surface topography or terrain that shows the undulations. A DEM can be represented as a matrix of pixels, where the DN value of each pixel represents the height (e.g. in metres from the mean sea level). Alternatively, it can be represented as a triangular irregular network (i.e. irregularly distributed nodes and lines with three dimensional coordinates that are arranged as a network of non-overlapping triangles). Unlike imaging sensors discussed so far, the raw data captured by these sensors is not always in raster format.

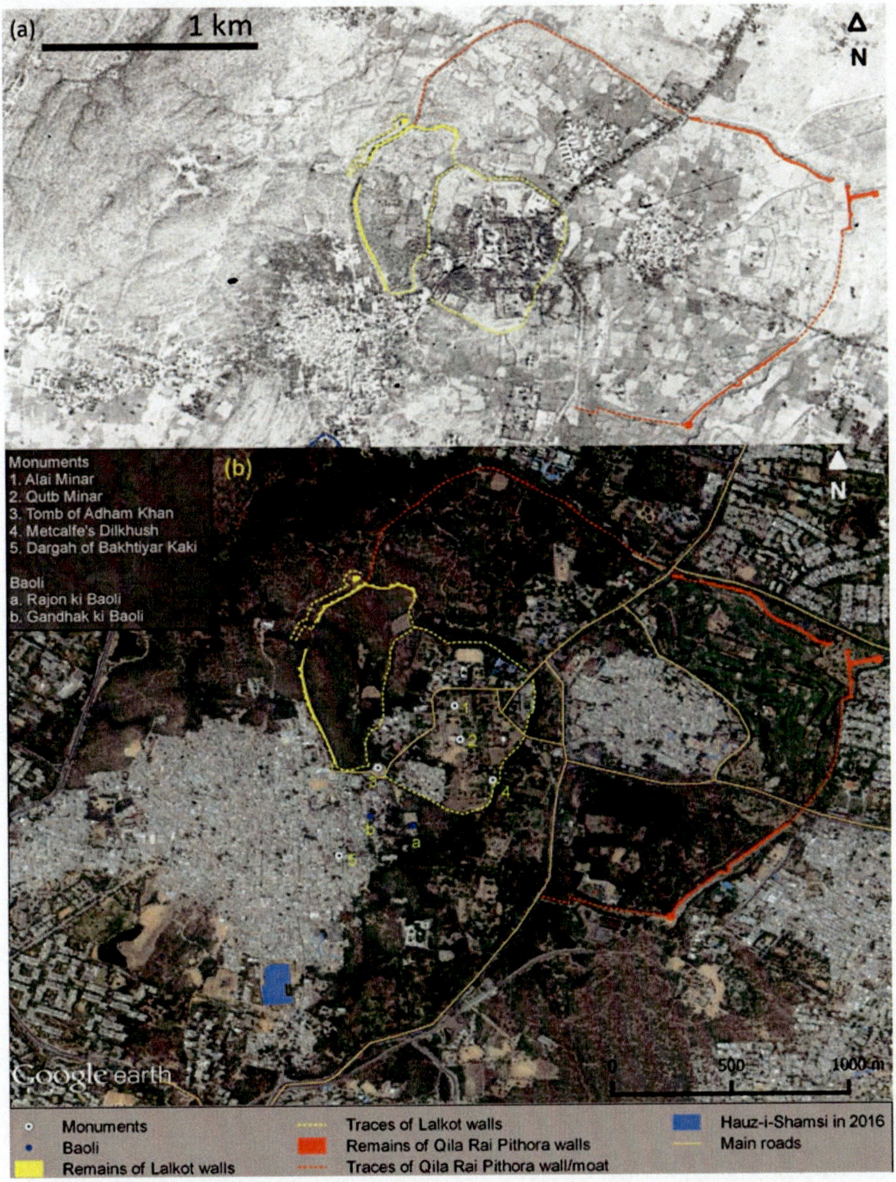

Fig. 3.13 Fortification of Lalkot and Qila Rai Pithora in Mehrauli, New Delhi, as seen on **a** Corona image of 8 October 1965; **b** Google Earth 5 October 2016

Fig. 3.14 Aerial photograph of Purana Qila taken on 15 February 1946, from Bradford Papers, Pitt Rivers Museum, Oxford

For instance, LiDAR sensors on aerial platforms scan the terrain with return laser beams that produce measurements with 3D coordinates *x*, *y*, and *z* called a point cloud (see Sect. 3.3.3). This 3D point cloud must be processed further to create a DEM in raster format. (The denser the return beams, the higher the spatial resolution of the DEM). In contrast, stereoscopic DEM is generated using stereoscopic images which are in raster format. (A pair of images of 2.5 m spatial resolution can be processed to produce a DEM of resolution of about 4–5 m.)

Such digital topographical data is not directly captured by sensors but are generated in several ways, and the most common ways are explained below.

3.3.1 Space Stereoscopy

Photographs of an area are taken from two different locations by optical sensors to determine height information using the principles of binocular vision. For aerial imaging, photographs are taken along a flight path with about 60% of overlap. Thereafter, the height information of each point in the overlapping region can be determined using parallax measurements through a stereoscope. To accurately determine height, the ratio of the distance between the two points where the photographs were taken (known as the base *B*) and the aerial platform's height (*H*), known as the *B*/*H* ratio, should be greater than 0.5.

For space-based imaging, there are two ways in which stereoscopic imaging can be realized: across-track and along-track. In across-track stereo imaging, images are acquired from two different orbits (possibly even from two different sensors/satellites). Generally, it is found that such stereoscopic views are not ideal because of varied illumination and surface conditions due to the time lapse between the two observations. Also, the spatial resolution from the two orbits can vary, and the *B*/*H* ratio is usually lower than 0.5. It is also possible to acquire the two images by tilting the mirror in front of the optics of the sensor, as in case of SPOT and IRS-1C and 1D satellites. Along-track stereo imaging comprises two or three camera systems at different viewing angles with respect to the nadir. Images are acquired at near simultaneous times and are designed to have a *B*/*H* ratio greater than 0.5. Cartosat-1 had two cameras at +26° and −5°, allowing a *B*/*H* ratio of about 0.65, with a spatial resolution of 2.5 m and a 30 km swath. These methods have been compared using stereo images of archaeological landscape of Badami (Rajani et al. 2009) and along-track is generally preferred (Fig. 3.15). A DEM generated using Cartosat-1 stereo images was used to identify the anthropogenic shape of Begampur (north of Nalanda), and

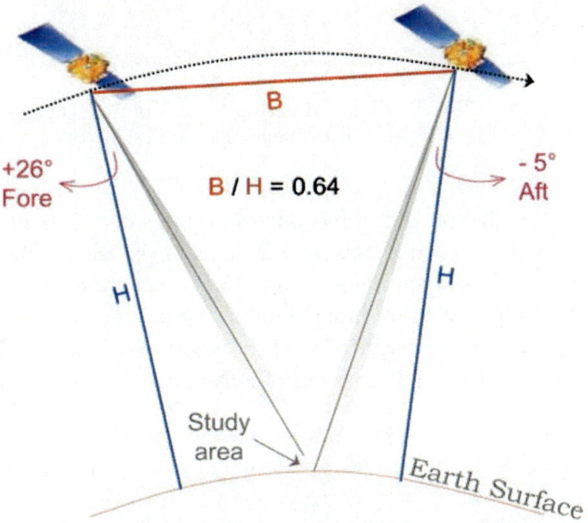

Fig. 3.15 Along track stereoscopy

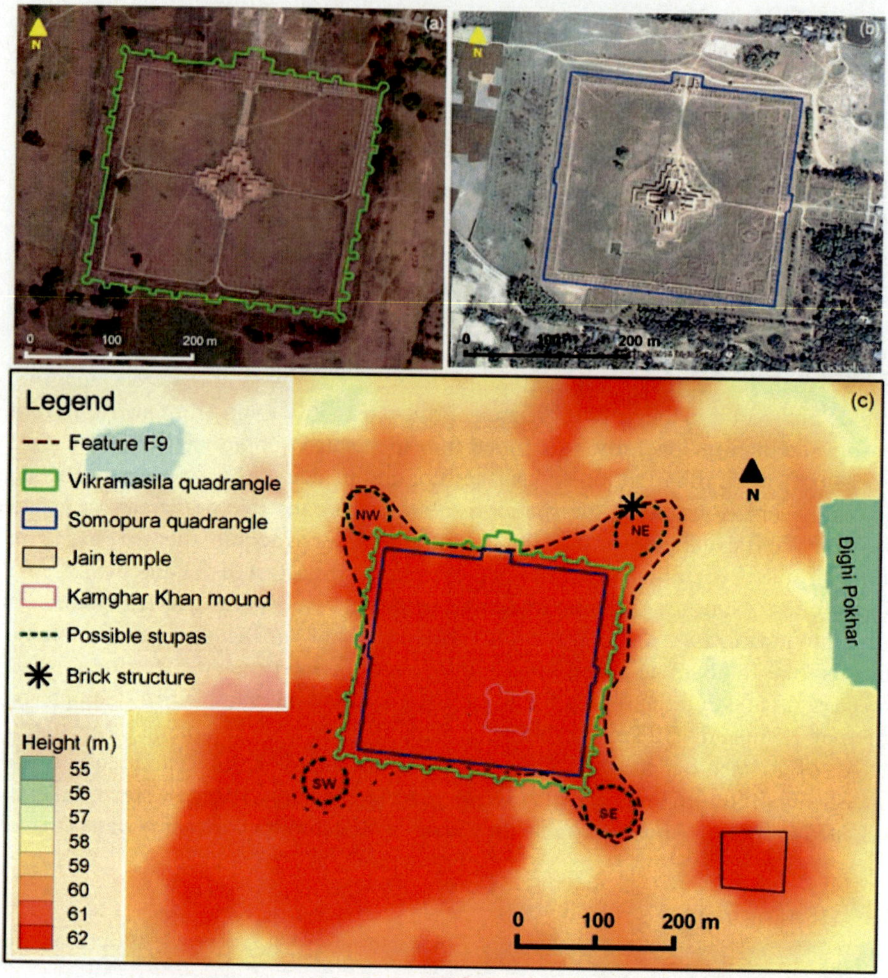

Fig. 3.16 **a** Vikramasila (in Bihar). **b** Somapura (in Bangladesh). **c** Begampur mound north of excavated site of Nalanda with quadrangles of (**a**) and (**b**) superimposed

we hypothesize that it has buried remains of a Pala period monastery that is of the scale in size and shape of Vikramasila (in Bihar) and Somapura (in Bangladesh) (Fig. 3.16) (Rajani 2016). Using Cartosat-1 stereo and photogrammetric techniques, a DEM for the entire Indian landmass has been generated at 10 m spatial resolution.[3] Efforts are on to create a DEM for many areas across the globe at similar resolutions. (The ASTER satellite system freely provides a 30 m spatial resolution.)

[3]https://www.nrsc.gov.in/sites/default/files/pdf/cartodem_bro_final.pdf. Accessed 11 May 2020.

3.3.2 Microwave

Microwave radar (i.e. SAR in interferometric mode) can also be used to generate digital elevation models of the earth's surface. Return pulses over a region obtained by SAR from two different locations in space (e.g. from adjacent orbits) or from two well-separated antennae on a single spacecraft differ in phase. Measuring and processing this phase difference provides very high-resolution digital elevation information. Shuttle Radar Terrain Mission (SRTM) has provided global 90–30 m DEMs using interferometric principles. The ASTER and SRTM DEMs are regularly used for archaeological and other applications, as they are freely available online. However, they are much coarser resolution (90 or 30 m per pixel) than what can be achieved by LiDAR scanning.

3.3.3 LiDAR

Terrain height can also be measured by light detection and ranging (LiDAR) sensors aboard aircrafts, UAVs and space-borne platforms (e.g. GLAS on board ICESat). LiDAR is an active sensor. A laser beam transmits several pulses per second to the earth's surface and measures the time taken by the reflected signal to return to the sensor. The position in terms of the X, Y, and Z coordinates of each measurement is obtained using inbuilt GNSS (Global Navigation Satellite System). This dataset consists of millions of measurements, called a point cloud. Analysing this point cloud using photogrammetric techniques provides the three-dimensional surface of the terrain, called a DTM. Since laser pulses have submicrometer wavelengths, the DTM obtained is of very high resolution. Such models have been used to reveal ancient cities buried under thick forests around Angkor Wat in Cambodia (Stone 2013) and Caracol in Belize (Chase et al. 2010). Terrestrial laser scanning is also gaining importance in many archaeological studies, where one can create a point cloud dataset for any monument or group of structures. If this is performed from multiple directions, high-fidelity three-dimensional models can be obtained for documentation and reconstruction (Prithviraj et al. 2012). These digital models can be 3D printed as scaled miniature models which are popular museum display material. An exhibition in January–February 2020 in National Museum, New Delhi, had riveting multimedia 3D displays of several Indian Heritage monuments.[4]

[4]http://www.ihse-event.org/2020/exhibition.php. Accessed 11 May 2020.

3.4 Data Preprocessing, Availability, Accessibility, and Sources

There are large constellations of remote sensing satellites that are continuously acquiring images of the earth surface, and every year several new earth observation satellites are placed in orbit. Many of these provide near real-time imagery for users. Many sets of archived images taken in the past are also available. For archaeological research, near real-time imagery is not a crucial requirement unless one is dealing with sites struck by a current event such as a natural or anthropogenic disaster (see Sect. 2.1.1). Most of the examples presented in this book have used archived (not real-time) satellite imagery.

3.4.1 Image Preprocessing

The raw data received at the satellite ground station have errors caused by the sensors, the platform, the intervening atmosphere, and the data transmission and reception systems. A degree of preprocessing is therefore necessary to correct the image data for distortions or degradations that stem from the acquisition process. After restoring the image data to a more faithful representation of the original scene, the data products can be made available to users. Here, we discuss a few typical types of preprocessing that most images are subjected to. Image providers usually provide metadata with every image which describes the changes made to raw image data.

Radiometric correction: The raw image often lacks contrast and therefore features we want to identify are usually not conspicuous. The lack of contrast can be due to radiometric errors, which are introduced due to sensor characteristics, the intervening atmosphere, and noise introduced during signal generation, transmission/reception. The radiance measured by any given system over a given object is influenced by factors such as changes in scene illumination, atmospheric conditions, viewing geometry, and instrument response characteristics. The need to perform correction for any or all of these influences depends directly upon a particular application.

Geometric correction: The rotation of the earth in east–west direction is near perpendicular to the direction in which satellites orbit (north–south). Hence, the raw digital images have considerable geometric errors, so much so that they cannot be used directly as a map base without subsequent processing. The sources of distortions are variations in the altitude, attitude, and velocity of the sensor platform, panoramic distortion, earth's curvature, atmospheric refraction, relief displacement, and non-linearity in the sweep of a sensor's Instantaneous Field of View. The errors can be broadly classified as systematic or predictable, and random or unpredictable errors. Systematic distortions are well understood and easily corrected by applying formula derived by mathematically modelling the sources of distortions. One important example of systematic distortion is the eastward rotation of the earth beneath the satellite during imaging. This causes each optical sweep of the scanner to cover an area slightly to the west of the previous sweep. The skewed parallelogram appearance of multispectral satellite data is the result of this correction. The errors produced by platform characteristics such as altitude variation, attitude (pitch, roll, yaw) disturbance and orbit drift, produce various geometric distortions random in nature, and cannot be easily modelled to correct images with high degree of geometric precision. Random distortions are difficult to eliminate fully but corrections are attempted by analysing well-distributed ground control points (GCP) occurring in the image.

Georeferencing: The data generated by remote sensors are images of the earth's surface, but the image pixels are not associated with specific locations on earth. Georeferencing accomplishes precisely this task, by transforming the image coordinate system (the column and row numbers of each pixel) to a particular map coordinate system. The map coordinate system is expressed in terms of latitude and longitude. Positional accuracy can be obtained from using GCPs in order to generate an accurate transformation model. A GCP is a feature which can be uniquely identified in an image and whose geographic coordinates are known or can be determined. The chosen GCP should be prominent, stable, and preferably a permanent feature that is not subject to seasonal changes so that its appearance is unambiguous. The coordinates of GCPs can be found out from existing maps or GPS surveys.

3.4.2 Image Sources

Although remote sensing first began for military purposes and use of these images were restricted, earth observation images have been increasingly used for many non-military applications. Therefore, even data that once highly classified (e.g. Corona images) is now freely available. Given the rate at which data is being gathered, and the pace of improvements in associated technologies that are constantly improving the resolutions and accuracy of data, any list of available sensors one makes is bound to become outdated. In the 1980s, for instance, spatial resolutions of 30 m per pixel may have been considered "high resolution", which they were compared with the coarser resolutions available at the time (50 m per pixel or more). Likewise, in 1990s, "high resolution" was understood as 5–10 m per pixel. The present yardstick for "high resolution" is better than 0.5 m per pixel. Table 3.1 lists various kinds of satellite images available at present, together with their characteristics and names of sensors. One can find information on the internet about image providers/distributors for any specific satellite/sensor images.

References

Belli FE, Koch M (2007) Remote sensing and GIS analysis of a Maya city and its landscape: Holmul, Guetemala. Wiseman J, El-Baz F (eds) Remote sensing in archaeology. Springer, New York

Bradford J (1957) Ancient landscapes: studies in field archaeology. G. Bell and Sons Ltd, London

Chase AF, Chase DZ, Weishampel JF (2010) Lasers in the Jungle. Archaeology 63(4) https://archive.archaeology.org/1007/etc/caracol.html. Accessed 13 Apr 2020

Lavers C (2019) Reeds introductions: principles of earth observation for marine engineering applications. Bloomsbury Publishing PLC. London, United Kingdom

Galiatsatos N (2004) Assessment of the CORONA series of satellite imagery for landscape archaeology: a case study from the Orontes Valley, Syria. Theses, Durham University. http://etheses.dur.ac.uk/281/. Accessed 13 Apr 2020

GEOBIA. https://clarklabs.org/segmentation-and-segment-based-classification/. Accessed 12 Apr 2020

Gibson PJ, Power CH (2000) Introductory remote sensing: digital image processing and applications. Routledge, London

Guio AD, Magnini L, Bettineschi C (2015) GeOBIA approaches to remote sensing of Fossil landscapes: two case studies from Northern Italy. In: 41st conference on computer applications and quantitative methods in archaeology. Perth, 25–28 Mar 2013, 2015, pp 45–53

Gupta E, Das S, Rajani MB (2017) Archaeological exploration in Srirangapatna and its environ through remote sensing analysis. J Indian Soc Remote Sens Springer 45:1057–1063. https://doi.org/10.1007/s12524-017-0659-9

Jackson VH (ed) (1922) Journal of Francis Buchanan (Patna and Gaya Districts). Journal of the Bihar and Orissa Research Society VIII(III & IV):263

Jesse C (2014) Regional-scale archaeological remote sensing in the age of big data: automated site vs. brute method. Adv Archaeolog Pract 2(3):222–233

Joseph G, Jeganathan C (2018) Fundamentals of remote sensing, 3rd edn. Universities Press Pvt Ltd, Hyderabad

Joesph G, Navalgund RR (1991) Remote sensing—physical basis and its evolution. National Academy of Sciences, India, Allahabad, 1991, published by M/S Malhotra Publishing House, pp 357–383

Kennedy D (1998) Declassified satellite photographs and archaeology in the Middle East: case studies from Turkey. Antiquity 72(277):553–561

McCauley J, Schaber G, Breed C, Grolier M, Haynes C, Issawi B, Elachi C, Blom R (1982) Subsurface valleys and geoarchaeology of the Eastern Sahara revealed by Shuttle Radar. Science 218(4576):1004–1020. Retrieved 12 Apr 2020, from www.jstor.org/stable/1688710

Navalgund RR, Rajani MB (2017) The science behind Archaeological Signatures from Space. Navalgund RR, Korisettar R (eds) Geospatial techniques in archaeology. Curr Sci Special Section 113(10)Nov 25th:1859–1872

Parcak SH (2009) Satellite remote sensing for archeology. London and New York, Routledge, Taylor and Francis group

Patil DR (1963) Antiquarian remains in Bihar. Kashi Prasad Jayaswal Research Institute, Patna

Prabhakar VN, Korisettar R (2017) Ground survey to aerial survey: methods and best practices in systematic archaeological explorations and excavations. Navalgund RR, Korisettar R (eds) Geospatial techniques in archaeology. Curr Sci Special Section 113(10)Nov 25th:1873–1890

Prithviraj M, Vijay UT, Kumar GCA (2012) Geo-spatial data generation and terrestrial scanning for 3D reconstruction. Int J Adv Res Comput Commun Eng 1(9)Nov

Rajani MB (2016) The expanse of archaeological remains at Nalanda: a study using remote sensing and GIS. Archives of Asian Art, Duke University Press. 66(1)Spring:1–23

Rajani MB, Kasturirangan K (2014) Multispectral remote sensing data analysis and application for detecting moats around medieval settlements in South India. J Indian Soc Remote Sens 42:651–657. https://doi.org/10.1007/s12524-013-0346-4

Rajani MB, Kumar V (2020) Where was Odantapuri located? Archaeological evidence using Remote Sensing, GIS and Photogrammetry. Resonance. Accepted, to be published

Rajani MB, Rajawat AS (2011) Potential of satellite based sensors for studying distribution of archaeological sites along palaeo channels: Harappan sites a case study. J Archaeolog Sci Elsevier 38(9) Sept:2010–2016. https://doi.org/10.1016/j.jas.2010.08.008

Rajani MB, Patra SK, Verma M (2009) Space observation for generating 3D perspective views and its implication to the study of the archaeological site of Badami in India. J Cult Herit Elsevier 10(1)Dec:e20-e26. https://doi.org/10.1016/j.culher.2009.08.003

Rajani MB, Bhattacharya S, Rajawat AS (2011) Synergistic application of optical and radar data for archaeological exploration in the Talakadu Region, Karnataka. J Indian Soc Remote Sens Springer 39(4)Dec:519–527. https://doi.org/10.1007/s12524-011-0102-6

Rajawat AS, Verma PL, Nayak S (2003) Reconstruction of Palaeodrainage NETWORK in North-west India: retrospect and prospects if remote sensing based studies. Proc Indian Nat Sci Acad 69A(2)March:217–230

Sabins FF, (1997) Remote sensing principles and interpretation. W.H.Freeman and Company/Floyd F Sabins

Stone R (2013) The Hidden City of Angkor Wat. https://www.sciencemag.org/news/2013/06/hidden-city-angkor-wat#. Accessed 13 Apr 2020

Wilson DR (2000) Air photo interpretation for archaeologists. Tempus, Gloucestershire

Chapter 4
GIS: An Array of Tools for Archaeology

Preamble

A geographic information system (GIS) is a software system that provides a framework for gathering, integrating, managing, and analysing spatial data. As we have noted earlier, it is often useful to look at multiple satellite images for a given site. These could be images taken by the same sensor on different dates (see Sect. 2.3.6), or from different sensors (see Sect. 3.2.1). GIS software organizes information into layers based on geospatial location, allowing the analyst to integrate data from multiple images, as well as spatial data from other sources such as ground truthing,[1] field observations, old maps, and other historical spatial records. In order to integrate spatial data, it is obviously necessary to ensure that all this data is *georeferenced* (i.e. it is expressed in a common geospatial coordinate system) so that features in one layer are as accurately aligned as possible to their footprint in other layers (Fig. 4.1). Thus, any geospatial analyst must be proficient in georeferencing. In the first part of this chapter (Sect. 4.1), we will discuss *why* georeferencing can greatly benefit archaeological and cultural heritage research. In the second part, we will discuss *how* to georeference the type of spatial data we typically encounter in this domain. We will split this discussion into three parts: georeferencing satellite images (Sect. 4.2), georeferencing historical spatial records (Sect. 4.3), and discussing other kinds of spatial analysis that adds value to research in cultural landscapes (Sects. 4.4, 4.5 and 4.6).

[1]By ground truthing we mean making direct observation to verify a feature already detected in satellite imagery, whereas field observation includes observing features afresh on the ground (not necessarily as an act of verifying a predetermined feature).

M. B. Rajani, *Patterns in Past Settlements: Geospatial Analysis of Imprints of Cultural Heritage on Landscapes*, Springer Remote Sensing/Photogrammetry,
https://doi.org/10.1007/978-981-15-7466-5_4

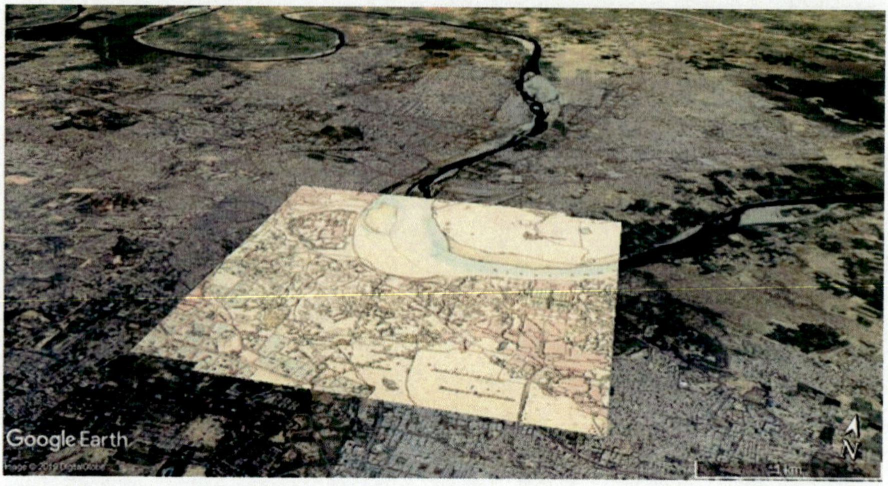

Fig. 4.1 Georeferenced Survey of India map overlaid over a satellite image. The spatial alignment between the two layers can be seen in the shape of the river

4.1 Why Is Georeferencing Useful?

An immediate benefit of georeferencing spatial data is that it facilitates precision measurements. However, the accuracy of these measurements depends on the accuracy of georeferencing and the spatial resolution of the images being used. In the context of studying past settlement patterns, typical measurements of interest include: the latitude and longitude of a specific point of interest (e.g. one that has been or will be investigated at ground level during a site visit), the dimensions of a linear or curvilinear feature (e.g. length of a wall or a street), the area of a feature with well-defined boundaries (e.g. a fort or a tank), and the orientation of a structure or a layout (either relative to other features, or to fixed directions such as geographical north).

Another benefit of georeferencing, particularly for large archaeological explorations or excavations, is that it provides a context in which the location of all material found at the site can be recorded. This systematic recording of spatial metadata is called *geotagging*, and it is a valuable book-keeping procedure (see Sect. 4.4). For example, for the nineteenth-century CE excavations, at the Buddhist monastery Nalanda in Bihar, there are no records of where many of the sculptures and other samples were found. Our study has identified several unprotected mounds in the vicinity of the protected site. We have collected a handful of brick samples from some of these mounds for carbon dating analysis, and each of these samples has been geotagged so that we can associate it with the structure it was part of (Das et al. 2019). Apart from physical materials, spatial metadata can also be gathered for other data collected at the site. For instance, Malik et al. have conducted ground penetrating radar (GPR) scanning of two sites within the fortified area of Vigukot in the Great Rann of Kachchh and have identified buried walls that are not visible on the surface at all (Malik et al. 2017). By georeferencing this data, the existence of these walls can be incorporated into the larger geospatial documentation of this site.

Finally, having a georeferenced backdrop of the site and integrating it with other geotagged artefacts and georeferenced layers allows the analyst to visualize some or all the layered spatial data in the form of visualizations such as maps and 3D views, to view the distribution of features within a site's spatial context (at various scales), to examine the spatial relation between specific features and propose hypotheses for such a relation. Such an integrated information system can assist an analyst investigating cultural heritage by shedding new light on well-studied problems and creating opportunities to ask new questions. We present some examples to illustrate the power of such visualizations. For instance, the density of the artefacts recovered from the archaeological excavation at the ancient Roman site Porolissum in present day Romania has been visualized as a heat map, which has enabled researchers to understand the spatial layout of human activity in the precinct (Opreanu and Lăzărescu 2015). Similarly, the locations of early, mature, and later period Harappan sites were visualized against a background of coarse-resolution multispectral satellite imagery. By doing so, we were immediately able to observe that mature period sites lay in dense clusters along the palaeochannels of what was once a major river (often referred as the River Sarasvati), whereas later period sites clustered closer to the Indus valley. This study used data from two independent sources, the first being georeferenced satellite image showing palaeochannels and the second being geotagged points from latitudes/longitudes available in published literature. As a result, we were able to propose a novel hypothesis: the palaeochannels were active during the mature Harappan period, but they subsequently dried and the lack of water forced later settlements to move closer to the Indus (Fig. 4.2) (Rajani and

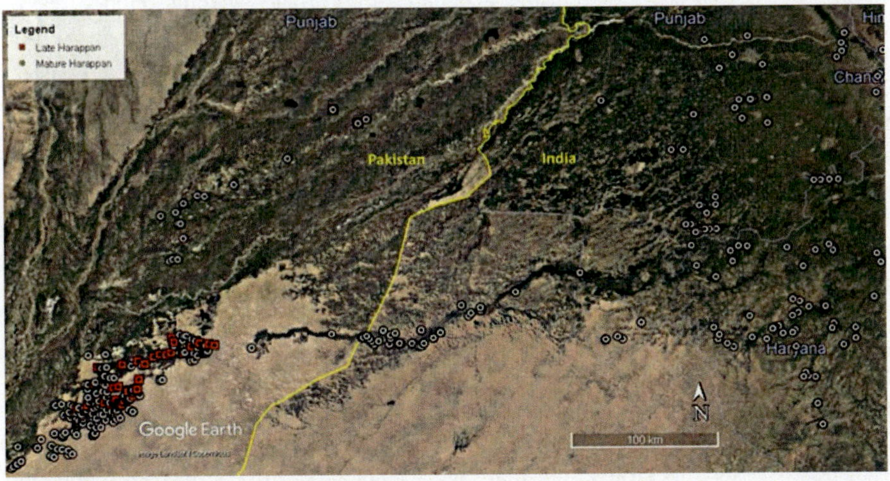

Fig. 4.2 Regional distribution of Mature Harappan and Late Harappan sites

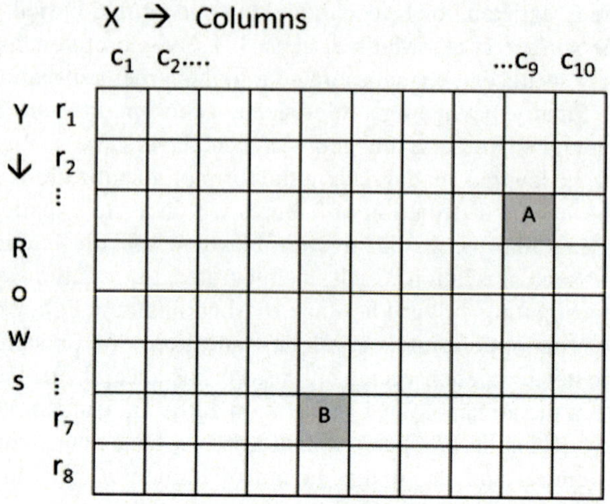

Fig. 4.3 Pixel A at column 9 row 3 and pixel B at column 5 row 7

Rajawat 2011). Similarly, by observing the distribution of manmade tanks and a palaeochannel which once drew water from a nearby river in Nalanda's spatial context (both identified on different images), we have hypothesized a much larger spatial extent for the Nalanda monastery as suggested by other historical records (Rajani 2016).

4.2 Georeferencing Satellite Images

As we shall see, there are multiple geospatial coordinate systems. Hence, even when certain spatial data such as modern satellite images are already georeferenced according to some coordinate system, they may need to be reprojected or even regeoreferenced in another system for compatibility with other layers of spatial information that the analyst wishes to examine. Older satellite images (e.g. Corona satellite images) must be georeferenced before they can be integrated.

An image is a matrix or *raster* of pixels. When a raw image is opened as a raster in GIS software, each pixel is assigned two coordinates: its column number and its row number (Fig. 4.3). The purpose of georeferencing is to transform the image coordinate system to a map coordinate system (i.e. each pixel will be assigned the latitude and longitude of the location it represents). This process requires the analyst to manually provide map coordinates for a small number of pixels (at least 4 to 5), known as ground control points (GCPs). The analyst chooses GCPs from features that can be uniquely identified in the *target* image (i.e. the image to be georeferenced) whose geographic coordinates are known or can be determined from another *source* such as an existing map, or ground-based GNSS measurements, or from geoportals like Google Earth, Google Map, Bhuvan, etc. A chosen GCP should be a sufficiently prominent feature that is not subject to seasonal changes so that it

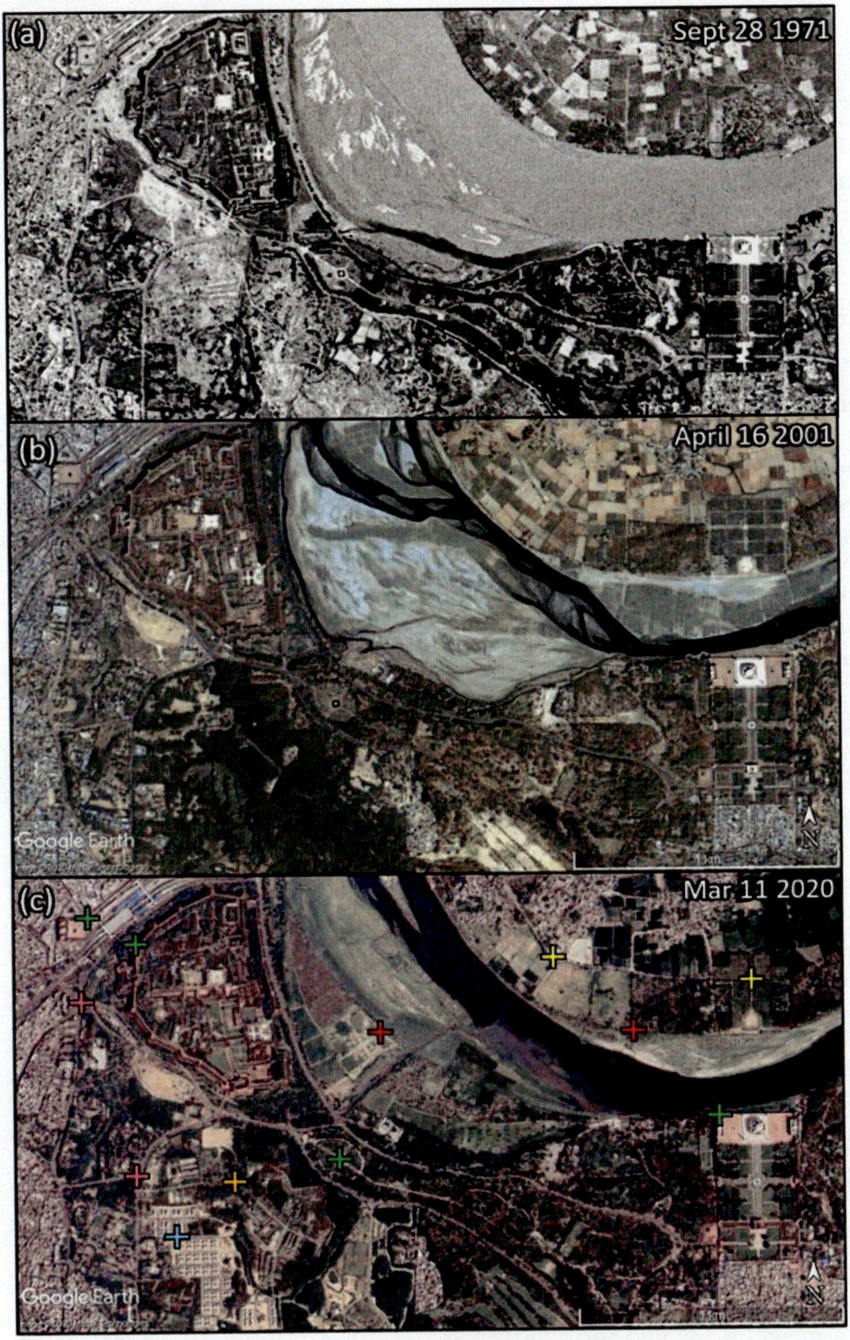

Fig. 4.4 Environs of Taj Mahal and the Fort at Agra **a** Corona satellite image; **b** and **c** images from two different dates, Google Earth. Several GCPs are marked in (**c**)

can be identified unambiguously in both the source and the target images. Further, it should be an immovable (preferably permanent) feature, at least over the period between the dates of these images. For instance, Fig. 4.4 shows three images covering the same area. Figure 4.4c has several points marked in crosses of different colour: the green crosses are well-defined corners of structures and pink crosses are road intersections. Both these features are visible in all three images. Hence, if any one of these was a source and the other two were target images, these points would serve as good GCPs. In contrast, the cyan cross represents a structure and the orange cross represents a road intersection that is only visible in Fig. 4.4c. Hence, these features are inappropriate as GCPs for the other two images, even though these are arguably "permanent" features. The yellow crosses represent features that are identifiable only in Fig. 4.4b, c, so these could be used as GCPs if either of these was the source and the other was the target image. However, these would not be useful GCPs if Fig. 4.4a was the target image. The red crosses are on the bank of a meandering river. Since rivers change course, such features are non-permanent features and are therefore not good GCPs. In general, it is better to have a small number of highly accurate GCPs than several inaccurate GCPs. Naturally, the larger the number of highly accurate GCPs, the better the positional accuracy.

After selecting several accurate GCPs, GIS software allows the analyst to assign each GCP a specific latitude and longitude and select references such as the coordinate system, datum, and projection. Given these assignments, the GIS software can automatically compute the latitude and longitude for the vast majority (often millions) of pixels according to one of several image transformations.

Since the earth is not a perfect sphere or even a perfect ellipsoid, there will always be some error in translating image coordinates to real-world coordinates. This error may be smaller for certain transformations. The simplest transformation is a first-order polynomial transformation, which can shift, scale, and rotate images. Linear features in the original image remain linear under such a transformation. If the original image is distorted in more complex ways (e.g. if it is a hand-drawn map), it may be necessary to use higher-order polynomials to correct for such distortions. In general, higher-order polynomial transformations require more GCPs.

Box 1 Coordinate system: coordinate systems enable layers of geographic data to use common locations for integration. A coordinate system is a reference system used to represent the locations of geographic features, imagery, and observations such as GNSS locations within a common geographic framework. There are two basic types: geographic and projected (Fig. Box 1a)

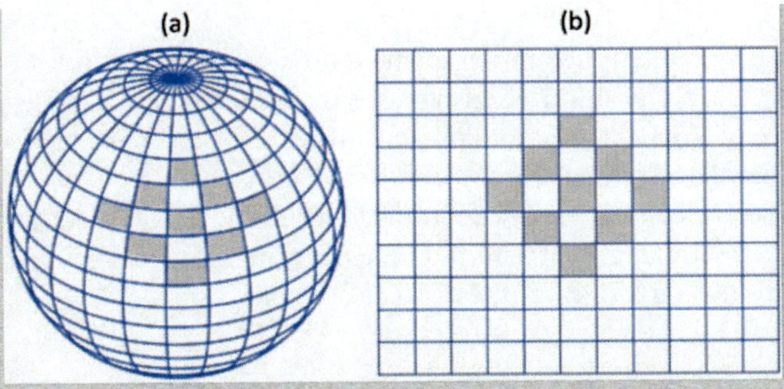

Fig. Box 1 **a** 3D Geographic and **b** 2D projected coordinate systems

. The former uses a three-dimensional spherical surface to define locations on the earth, where a point is referenced by its longitude and latitude values. The latter is defined on a flat and two-dimensional surface that is based on the grid of a spheroid or ellipsoid, where locations are identified by *x* and *y* coordinates on a grid, with the origin at the centre of the grid.

Box 2 Datum: a datum is a system which allows latitude, longitude, and height to be calculated for any location on the earth's surface. If the earth had been a perfect ellipsoid, this calculation would have been quite simple. Unfortunately, no ellipsoid model fits the earth precisely. Hence, geodesists have traditionally used several different ellipsoids, called local (or regional) datums, each of which is accurate only for certain regions. Figure Box 2

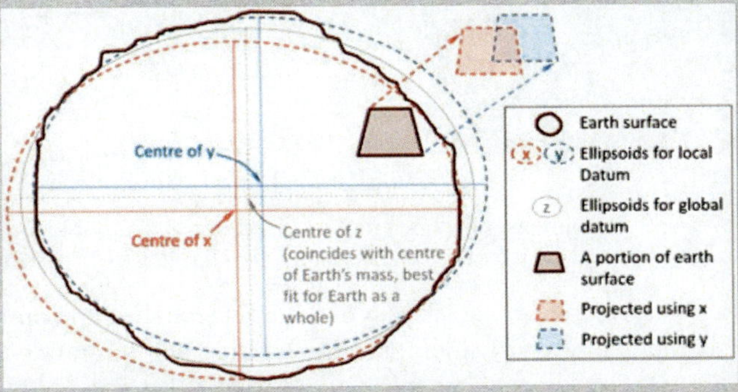

Fig. Box 2 Diagram showing global versus local ellipsoids. The misalignment between the red and the blue trapezia is due to differences between the two local datums

uses an exaggerated illustration to show how two different local datums *X* and *Y* align closely with different parts of the earth's surface. Outside these regions, two different local datums can disagree by over 100 m, which is unacceptable. Thus, global (or geocentric) datums such as World Geodetic System 1984 (WGS 84) have been developed (Fig. Box 2). As their name suggests, the centre of global datums coincides with the centre of earth's mass, and they provide a good fit for the earth as a whole.

Box 3 Projection: the earth is curved, whereas maps are flat (Fig. Box 3).

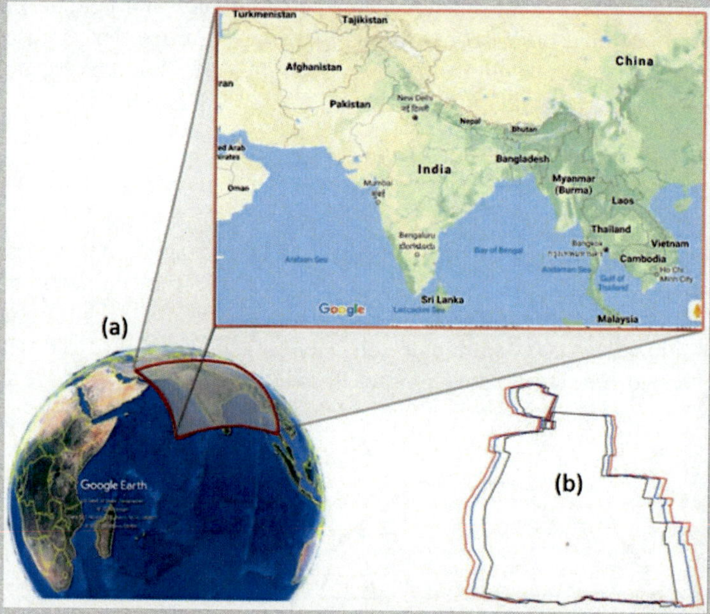

Fig. Box 3 Diagrams showing **a** projection from a curved to a flat surface; **b** differences between the boundary of Nalanda's core zone depending on the choice of projection

Projection is the mathematical process of flattening out the earth onto a piece of paper or computer screen, which uses the latitude and longitude "drawn" on the surface of the earth using a chosen datum. The illustration of Nalanda's core zone (discussed in Sect. 6.2.3) in Fig. Box 3 shows the distortions that can arise due to different projections in multiple layers.

4.3 Georeferencing Historical Spatial Records

In Sect. 4.1, we have discussed several advantages of georeferencing spatial information. For some sites, we are fortunate to have historical spatial records in the form of old maps, archaeological reports, travellers' records, literature, and epigraphy. While it is tempting to try and georeference all such records, this should be avoided when they lack spatial accuracy. In this section, we will discuss examples of records with sufficient accuracy to warrant georeferencing, how to georeference such records (when appropriate), and how to leverage spatial information from historical records with poor spatial consistency.

The idea of representing geographical environments in written forms dates to at least ancient Greece. While early maps were allegorical, the development of geometry had a significant impact on mapping. Greek scholars used geometric principles and techniques to determine the shape and size of the earth and to determine the relative positions of environmental features. Geometrical concepts also led to the development of locational reference systems, such as latitude/longitude. In classical cartography, the positions of stars were used as reference points (Berggren and Jones 2000). The next major revolution in cartography was the trigonometrical survey, where elevated locations on the ground were used as reference points (Keay 2000). Modern mapping technology such as GPS uses the location of satellites relative to specific reference points on earth to locate positions and measure distances (Williams 1994). There were very few accurate copies of early maps (and hence many are lost forever). By the 1900s, however, the development of printing and photography made reproduction far easier and cheaper.

While analysing historical maps, it is necessary to be aware of the methods used in creating them, because each of these methods has limitations that must be accounted for during georeferencing. Maps are historical documents and the standards for producing them (such as what features to include—or even exaggerate—and what to ignore) have evolved over time. Such inconsistencies can pose a challenge (Gupta and Rajani 2020a).

4.3.1 Maps Made by Trigonometrical Surveys

Among historical maps, those made using trigonometrical survey and vertical orthographic projection (top-down view) are generally the most spatially accurate: the locations, shapes, and proportions of features they depict are often consistent with identifiable features in georeferenced satellite imagery.

Triangulation was used as a method for drawing local maps to scale in Europe as early as the fifteenth-century CE (Harvey 1987). In 1533, the Dutch mathematician Gemma Frisius published a booklet proposing triangulation as a method for making regional maps. This idea was implemented by cartographers across the Netherlands, Germany, Austria, and England (Andrews 2009), and Willebrord Snell, another

Dutch mathematician, gave us the modern systematic use of triangulation networks (Kirby 1990). By the middle of the seventeenth-century CE, the first such maps for sites in India were created by the Dutch East India Company for their colonies, which included Cananor (Kannur), Cranganor (Fort of Kottupuram south of Kodungalur), and Coylan (Kollam) along the coast of modern-day Kerala (Asia Maior 2006). Over the next century, cartography evolved from producing local maps—including maps of cities and towns of India such as Masulipatnam (Beveridge 1900) and Madurai (Jennings 1755a, b), see Figs. 2.14 and 2.19, by the British and of Pondicherry (*Plan de Pondicherry* 1741) by the French—to maps of whole countries, including France (1745), Great Britain and Ireland (1783 to 1853), and India (1802 to 1871) (Andrews 2009). The latter study, known as the Great Trigonometric Survey of India, served not only the political need of demarcating British territories but also advanced scientific understanding by measuring the height of several Himalayan mountains (including Mt. Everest), and advancing our knowledge of the shape of our planet by carefully studying the earth's curvature along meridian arcs (Keay 2000).

Countries started establishing agencies to standardize mapping practices, such as the Survey of India (SOI) in 1767, the US Geodetic Survey, 1807, Principal Triangulation of Britain began an in 1791. These standards evolved in several stages before modern digital maps became the norm. For instance, the shapes of features such as roads, settlement layouts and fort walls in maps made under the aegis of Survey of India can be considered spatially accurate from the nineteenth century onwards, even though standards continued to evolve well into the next century. It is important to keep this in mind when examining maps from this period, as the following examples will show. In 1870, the SOI published a map of Agra, the city that is home to the Taj Mahal, which marks several features with high spatial accuracy, including the river, ravines, ditches, monuments, and localities. However, the map does not mark the two-tiered city walls, which other sources confirm were present at the time (Suganya and Rajani 2020). Similarly, an early twentieth-century SOI map of Bangalore marks *karez*. This terminology, which is of Persian origin, suggests underground tunnels to supply water. However, there are no references to *karez* in the textual literature of this region, or even in other maps of Bangalore from the late eighteenth, nineteenth, and early twentieth centuries. Our investigation into this inconsistency has revealed that, at the time, SOI used the term *karez* to mark any gravity-based underground water porting system (Suganya and Rajani 2018).

Revenue survey maps of India to meet the needs of constructing canals, railways, roads, etc., were separately published (Thuillier and Symth 1875), but in 1878, the SOI took over responsibility for surveying and publishing all maps of India for both civilian and military purposes (Wheeler 1955). The *Manual of Surveying for India…* laid out the rules regarding drawing, finishing, and publishing of maps, stating that all the elements on ground should be marked neatly and clearly, and the symbol should represent and imitate the form and appearance at first look (Thuillier and Symth 1875). Military maps would mark all means of communication (such as railway lines, roads, trails, paths, streams, rivers, and canals), types of shelter for the troops (cities, towns, villages, and isolated houses), sources of water supply (smaller streams, lakes, ponds, large springs, and wells), sources of fuel (woods, orchards, crops, and grass land), and obstacles to the movement of the troops and form of the ground that are required by the army (Thuillier and Symth 1875; Stuart 1918). Thus, it stated that maps published by the SOI should depict accurate and clear topographical details of all natural, economical, built, transportation, and communication features (Thuillier and Symth 1875; Wheeler 1885).

However, in 1905, a committee consisting officers from civil, military, and the SOI along with advisors from Ordnance Survey of Great Britain reviewed existing maps of India from various circle offices and found that they were deficient in topographical details. In addition, there were inconsistencies in the variety of information they contained and the way in which this information was depicted (Longe 1906). The committee drafted policy and guidelines for rectifying the same (Wheeler 1955). The policy recommended preparing a series of colour maps for the whole of the Indian subcontinent in "one inch to a mile" (1:63,360) scale within 25 years, marking heights in intervals of 50 feet, and using a uniform system of symbols for the whole of India, Baluchistan, North-West Frontier Province, and the adjacent countries (Longe 1906). This system was followed, with minor modifications, until WWII (Wheeler 1955). After India's independence, the SOI catered to the mapping required to meet the needs of defence, planners, scientists, and for land and resource management. It produced maps in 1:250,000, 1:50,000, and 1:25,000 scales.[2] From the 1980s onward, the SOI has ventured into the phase of generating a digital topographical database for the entire country for use in various planning processes and the creation of geographic information systems.

4.3.2 *Eighteenth and Nineteenth-Century Maps Made Using Other Methods*

Many maps created in the Eighteenth and nineteenth century used a mixture of survey methods. For example, distances between important landmarks may have been measured using triangulation, whereas distances between less important features were recorded as estimated walking distances. These methods have very different margins of error, ranging from a few metres to several kilometres depending on the scale of the map (Hesse 2016). For this reason, such maps are often distorted by attempts to georeference them using standard methods that take all spatial information into account. For such maps, it is important to perform georeferencing using only the well-surveyed points. As an example, let us consider Alexander Cunningham's *Sketch of the ruins of Nalanda* from 1871 (Fig. 4.5a), one of several such site maps he made in the second half of the nineteenth century. The sizes of each archaeological mound marked on the map and the proportion of distances between them are to scale when compared with respective features identified on georeferenced DEM and Google Earth images (Fig. 4.5b, c). In contrast, the shapes and sizes of the water bodies—Gidi Pokhar, Pansokar Pokhar, and Indra Pokhar—and the roads/paths are

[2] http://www.surveyofindia.gov.in/pages/view/10-about-us. Accessed 12 May 2020.

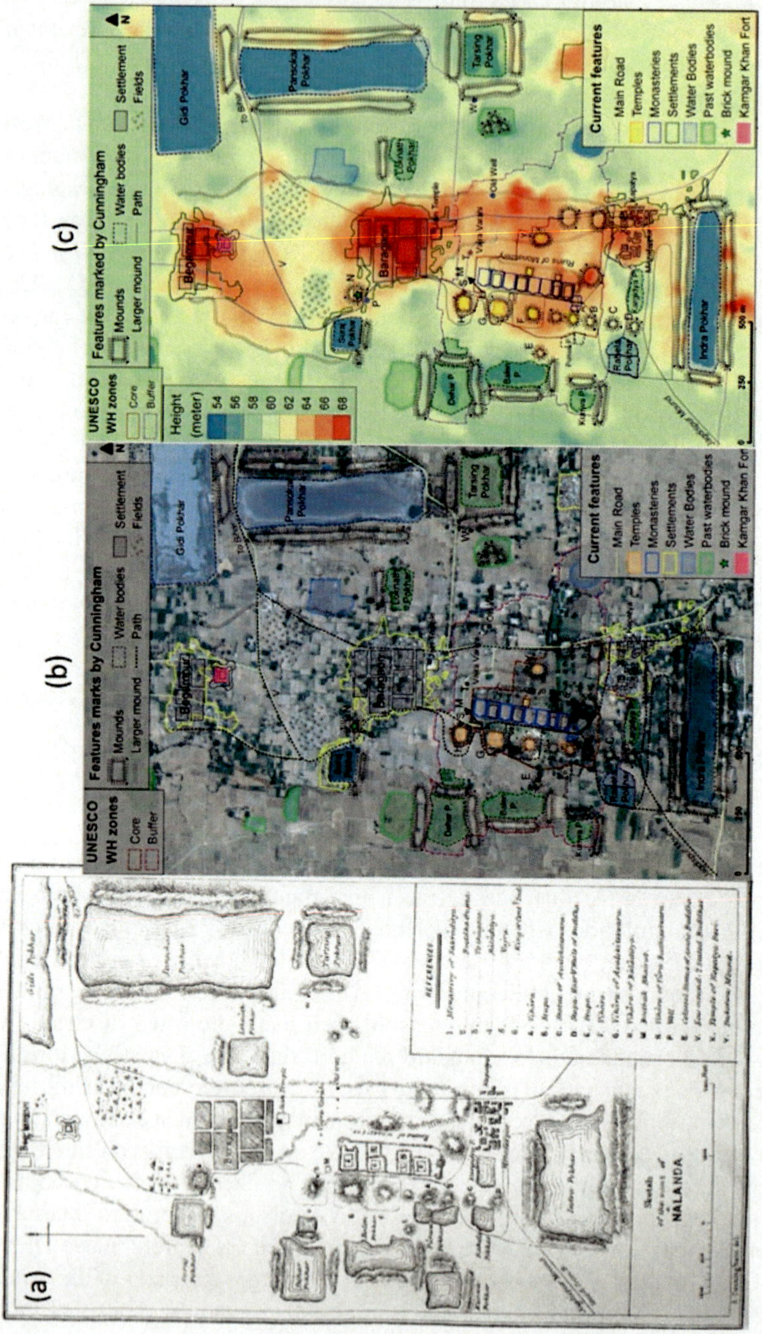

Fig. 4.5 **a** Sketch of the ruins of Nalanda (reproduced from Cunningham 1871, Plate XVI); **b** features from the sketch overlaid on a Google Earth image; **c** the same overlaid on a DEM

filled in the interim spaces. Thus, if we georeference this map using archaeological mounds as GCPs, we will find far less distortion than if we used road intersections or the corners of water bodies. However, even the former approach is challenging since mounds are not one-point features. Instead, we suggest the following approach for such maps. We begin with a georeferenced satellite image of the site. Next, we identify a few reliable points from the map and mark these as a vector layer over the satellite image. (Hence, these points acquire coordinates from the underlying satellite image.) Finally, we use these as anchor points and trace the remaining features from the map by comparing them with visible features on the satellite image. Note that this can be a laborious manual process, but we have found the trade-off between effort and accuracy to be acceptable. In case of Cunningham's Nalanda sketch, we first recognized that the mounds he identifies as H, G, F, A, and Y correspond to the subsequently excavated temples 14, 13, 12, 3 and the Sarai Temple, respectively. We were then able to correlate features C, D, X, and N on Cunningham's sketch with unexcavated mounds identified in the vicinity of the site. Cunningham's features M and T correspond to shrines that are currently used for worship. Having fixed these points, we drew vector shapes to match their depictions in the sketch. Thereafter, the similarities in the arrangement of waterbodies, roads, and settlements became conspicuous. We traced these one by one.

This kind of spatial accuracy in archaeological features depicted in the site maps or sketches made by Cunningham is consistent across the sketches he made of various sites. Figure 4.6 of Sarnath is another example, where the location of Dhamekh, Dharmarajika, and Chaukhandi stupas marked as mounds in Cunningham's map and the location of Saranganatha temple was used to first fix the reference between the sketch and Google Earth imagery. With this in place, the shapes of water bodies and road became conspicuous. We were then able to mark several of the oblong mounds around the three connected water bodies. Finally, we were able to visit these locations for verification, and we found that several of these mounds still exist.

A large set of Cunningham's maps for sites across north and east India were published in the early reports of ASI. By understanding their limitations, choosing GCPs, and tracing features with care, valuable spatial information for these sites can be extracted from such maps if one is willing to invest in the manual effort.

4.3.3 Sea Charts and Maritime Maps

Portolan charts or sea charts also fall into the category of historical spatial records. These navigation maps were made in Europe from the thirteenth-century CE onwards because major trade routes used the sea, marine channels and navigable rivers. Portolan charts were based on realistic descriptions of harbours and coasts as suggested by the word *portolan*, which comes from the Italian adjective *portolano* meaning "related to ports or harbours" (Rajani and Kasturirangan 2013). These charts identify relatively permanent features such as forts, temples, and hills that were visible on the coast at the time along with coastal geomorphological features such

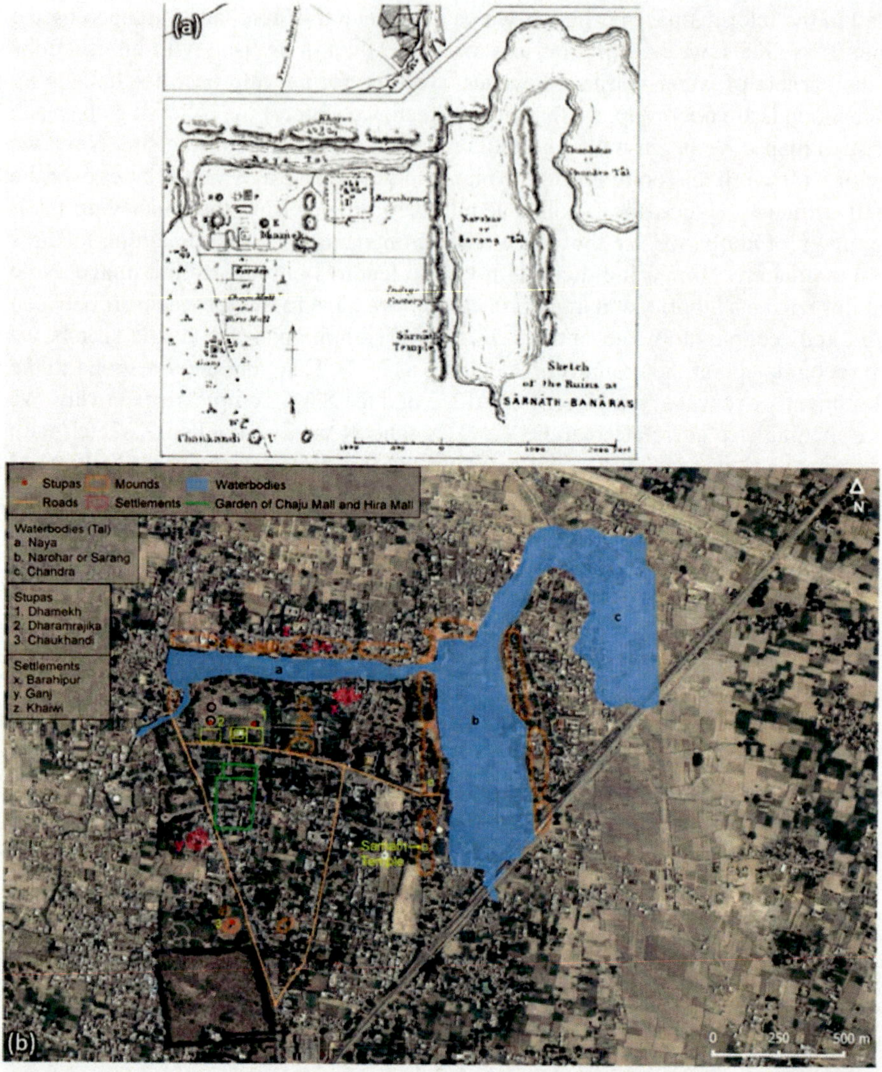

Fig. 4.6 **a** Cunningham's sketch of the ruins of Sarnath (reproduced from Cunningham 1871, Plate XXXI); **b** the same overlaid on a Google Earth image

as river-mouths, mud-flats, spits, and bays. An example of a Portolan chart of the seventeenth century of the present Tamil Nadu coast is shown in Fig. 4.7. The coastal structures marked on this chart are also visible on Google Earth and Bhuvan satellite imagery. Section 2.3.7 discusses a portion of this Portolan chart which shows details of Mahabalipuram which aided in understanding why the site had the toponym Seven Pagodas. These charts and the features marked in them (coastline, places, hills, buildings, and trees) are often not to scale. But these were made very systematically

Fig. 4.7 **a** Dutch Portolan chart (1670) from the archives of the Royal Geographical Society, London; **b** the corresponding coastal stretch on Google Earth marking a few common locations

with the then state-of-art techniques. We believe that these charts represent a largely untapped resource of immense value for studying coasts. One can derive more meaningful information from these charts if they are analysed in context of the status of knowledge in relevant fields at the time, and with awareness of the limitations in the methods used for creating them (Gupta and Rajani 2020a).

4.3.4 Paintings and Freehand Drawings

There are several historical records that present spatial information graphically, but in a manner that is too inaccurate for georeferencing. Nevertheless, for sites where no other forms of spatial records are available, it is sometimes possible to extract valuable information from such resources. The analysis tends to be very different for each site, but we can articulate two general principles. First, even if the analyst cannot trust the proportions and distances depicted in these maps or paintings, the *relative arrangements* of features (particularly neighbouring features) are often reliably captured. Second, features common to multiple historical spatial records can help establish valuable spatial information, even if these records are individually spatially inaccurate. We illustrate these principles with two examples that also highlight the site-specific nature of the analysis one must expect in settings with limited accuracy of spatial data.

Patna: among Francis Buchanan's diaries from the Bengal survey (1807–2014) is an unpublished map from 1812 of Patna, today the capital of Bihar state in India (Buchanan 1812) which was made by an assistant and has annotations in Persian (Jackson 1925). Patna is depicted as a linear settlement running along the bank of the River Ganga. The map marks a rectangular fort north-east of the settlement (see Fig. 4.8a), several roads, temples, and important landmarks. At ground level, it was difficult to discern the fort even when Buchanan visited. He records: "the fort in the north-east corner of the city is now so overrun with modern buildings that its form can be no longer distinguished, nor could I perceive any remains, except some old gates". Buchanan also noted an inscription on the fort wall attributing its erection to Feroz Jung Khan (Jackson 1925), although Cunningham later stated that the fort was built by Sher Shah Suri (a theory which supports Cunningham's argument that Patna was Palimbothra based on renovation of the old ruined fort by Sher Shah Suri) (Cunningham 1880). However, Cunningham does not provide a map or any other spatial evidence for where this fort was located. A comparison of the map (Fig. 4.8a) and the satellite image (Fig. 4.8b) reveals substantial changes to the land-cover. As a result of these changes, it is impossible to identify a sufficient number of GCPs to georeference this map.

In fact, the only structure we can identify with some certainty is Golghar (Brown 2005), a dome-shaped granary[3] built in 1784 that is maintained by the Directorate of Archaeology, Bihar (see Fig. 4.8c, e, f). Extracting further spatial information from such maps necessarily involves a degree of guesswork. On the satellite image, we can identify the location of a temple that bears the name *Paschim Darwaza* (western gateway). Although there are no remains of the fort in its vicinity today, we can surmise from the temple's name and from the location of the fort's western gate relative to the river and Golghar as depicted on the map that this temple was near the western gate. As noted in Sect. 2.3.3, roads tend to develop adjacent and parallel to fort walls, and we find such a road near the *Paschim Darwaza* temple leading to the river. In fact, we can identify roads that may have corresponded to all four sides of the rectangular fort (Fig. 4.8b). As further confirmation, the rectangular area within these roads is also slightly higher than the surrounding area (see the DEM in Fig. 4.9), consistent with the hypothesis that the development in this area sits atop the remains of a rectangular fort.

Thus, despite the lack of accurate spatial information in this map, it is valuable for analysis. Note that the map does have a rough scale along its edge (Fig. 4.8a, c), with the distance between evenly spaced cross marks subdivided into ten segments by vertical marks. The accuracy of this scale is questionable, and the map does not mention the unit of distance. However, thanks to the preceding analysis, it is possible to test the accuracy of this scale, for instance, by comparing ratios between the fort's length, breadth, and the distance between Golghar and the fort's western wall as

[3] http://yac.bih.nic.in/Da-02.htm#Golghar. Accessed 12 May 2020.

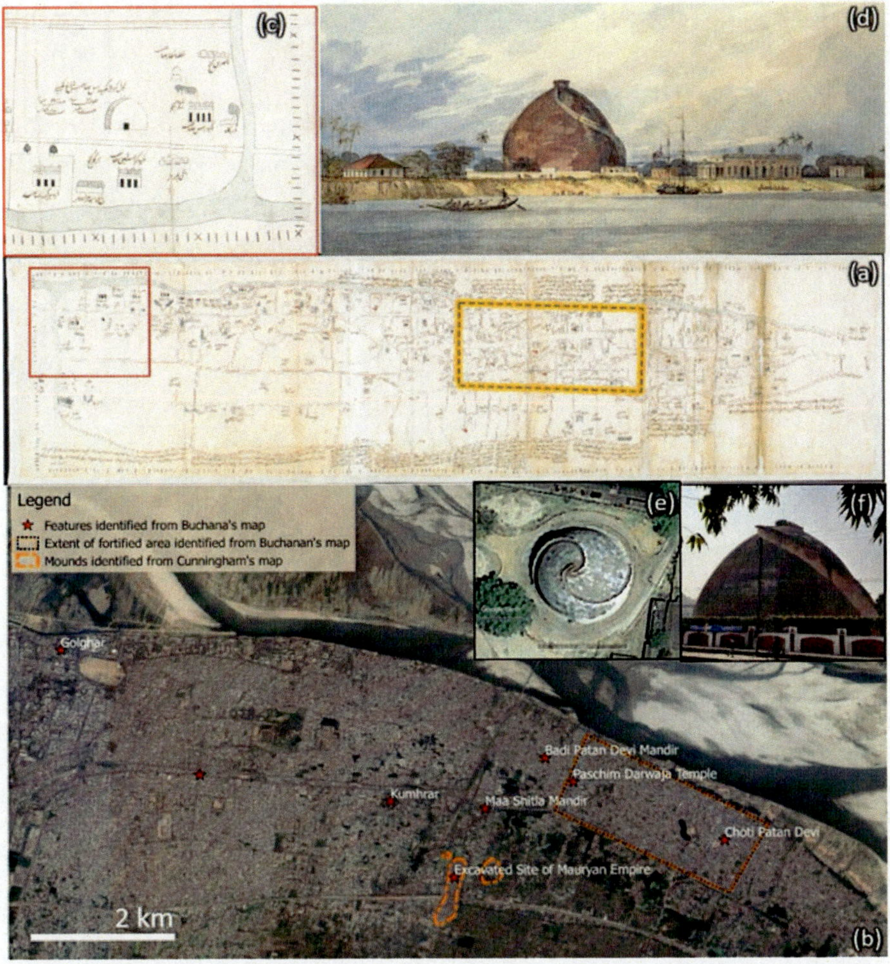

Fig. 4.8 **a** Plan of Patna from Buchanan's record (1811–2012) with the fort marked in yellow (© British Library Board; WD2090, 2090); **b** the corresponding region on Google Earth showing Patna and the features identifiable from the plan; **c** a portion of the plan showing Golghar; **d** a painting from 1814 showing a view of Golghar from the river; **e** Google Earth imagery of Golghar; **f** a recent ground photo of Golghar

represented on the map against the same ratios as measured on the georeferenced satellite image. If these ratios prove to be similar, our confidence in the spatial fidelity of this map would grow and could lead to further insights.

A painting of the Golghar (Smith 1814) dating from 1814 shows the view from the river as well as several other buildings around the domed structures which are also marked on the Persian map (Fig. 4.8d). Using these as pointers, we could identify the extent of Patna when Buchanan visited and also the fort marked in the map (Das and Rajni 2020).

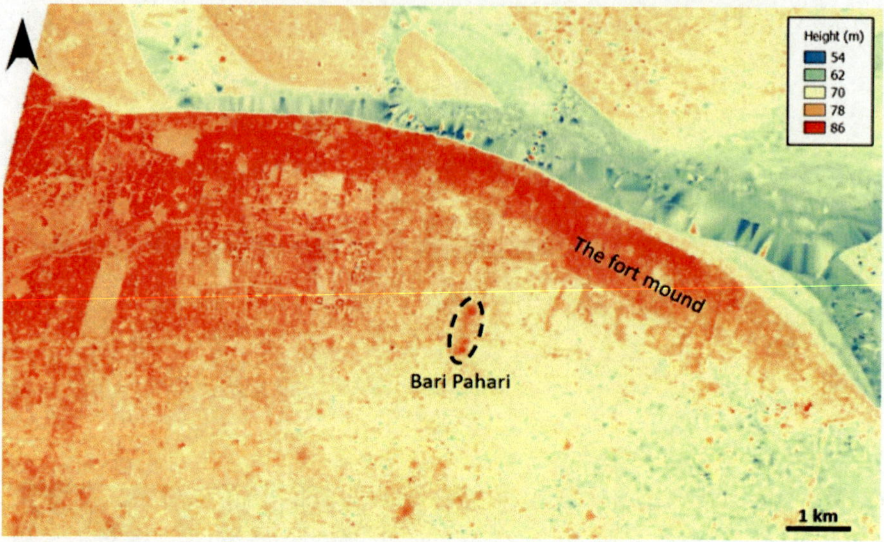

Fig. 4.9 DEM of Patna showing the elevated area of the rectangular fort

Kollam: two Portuguese atlases dating from 1630 and 1635 illustrate several port sites on India's west coast (Albernaz et al. 1630; Bocarro 1635). The forts and other buildings are depicted in an oblique or isometric view. Both these atlases have almost the same set of sites but the latter, probably used the former but is not a facsimile, and both are not to scale. Figure 4.10a shows the port of Kollam (*Covlao*) from the earlier atlas and Fig. 4.10b shows the same port (*Covlam*) from the later atlas. The former map includes details such as the title, labels, and direction, whereas the latter omits these but shows greater architectural details. Figure 4.10c is a pen and brush drawing entitled *View of the fort at Coilan* by Johannes Rach, made between 1760 to 1780. This drawing shows the bay as viewed from the east (facing west), centred on the three-storeyed structure that is also prominent in Fig. 4.10a, b. Based on this evidence, we hypothesize that this corresponds to the three-storeyed ruins of the central tower (Fig. 4.10d) of the Fort St. Thomas (Forte de São Tomé, also known as Tangasseri Fort) protected by the Archaeological Survey of India (Innes 1997). The location of this tower is marked with a red circle in the satellite image Fig. 4.10f. We also have a planimetric Dutch map made in 1687 by Hans Georg Taarant/Tarand and Laurens Nicolaesz Duyrendaal (Fig. 4.10e) (Taarant and Duyrendaal 1687).

Here again, the available historical spatial records are too inaccurate to georeference. However, since we know the coordinates of the three-storeyed structure in these figures, we can look for patterns in satellite imagery to identify the walls of the inner fort (Fig. 4.10e) and the outer fort (Fig. 4.10a, b). The moat surrounding the inner fort has left traces in the form of positive crop marks (Sect. 2.3.2), but these are challenging to identify because several modern buildings obscure the shape. However,

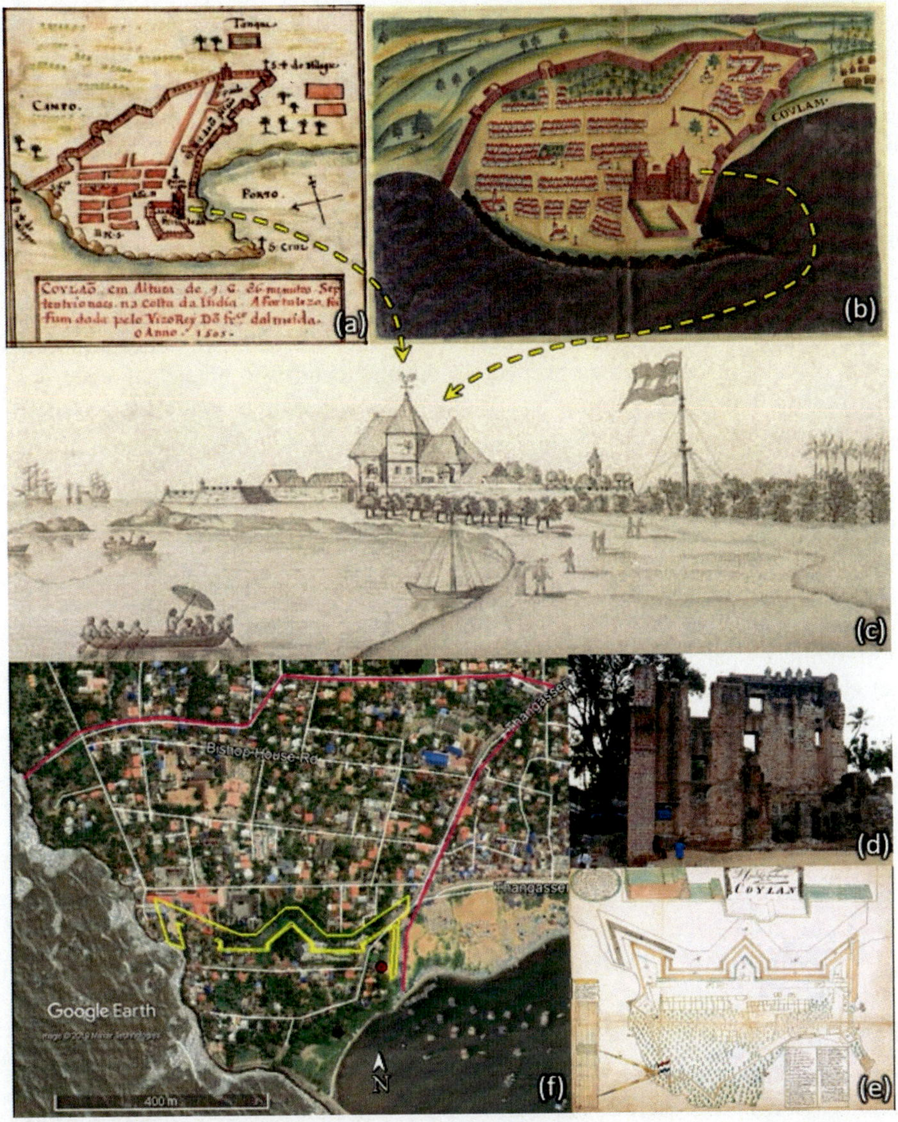

Fig. 4.10 **a** Map of *Covlao* in Albernaz et al. (1630) (Geography and Map Division, Library of Congress, USA); **b** a map of *Covlam* in Bocarro (1635) (Biblioteca Nacional, Portugal), **c** the drawing *View of the fort at Coilan* by Rach (1760–80) (National Library of Indonesia); **d** the ruins of Fort St. Thomas in Tangasseri, Kollam, photographed in Dec 2018; **e** a map of *Coylan* in Taarant and Duyrendaal (1687) (Nationaal Archief, Netherlands); **f** Google Earth, Maxar Technologies, 10 June 2018, the fort from **e** is marked in yellow, the wall from (**a**) and (**b**) is marked in pink, and the location of the structure in (**d**) is marked with a red dot

we were able to confirm our hypothesis when we visited these locations for ground-truthing and found evidence of ruins of the fort wall and adjacent depressions of the moat (Gupta and Rajani 2020b). We were unable to detect crop or soil marks corresponding to the walls of the outer fort. However, the pattern is well preserved in the roads (shown in pink) in Fig. 4.10f, including the distinctive V-shaped indentation seen in Fig. 4.10a, b. As noted in Sect. 2.3.3, a road may merely run parallel to (not on top of) former fort walls. Considering the proportions of distances from various parts of the inner fort, we believe that the roads highlighted in pink run outside the walls of the outer fort.

Our analysis of Kollam demonstrates that historical paintings such as Fig. 4.10c are valuable sources of spatial information. For sites in India, we believe that the wealth of paintings and drawings made in the eighteenth and nineteenth centuries CE (most famously by Thomas and William Daniell; Mitter 1977) are a largely untapped resource.

4.3.5 *Textual Sources*

We briefly note that valuable spatial information can sometimes be recorded in textual form. While textual descriptions of features rarely mention geographical coordinates, they can refer to nearby features whose locations can be ascertained. As an example, Buchanan noted in 1812 that to the west of a tank called Dhigi, on the outskirts of Nalanda, he saw a large mound with fragments of bricks. He added that on the north end of this mound was the village of Begampur (Rajani 2016). Based on this textual evidence, we have identified this feature as the squarish mound to the north of Nalanda seen in Fig. 3.16c.

4.4 Geotagging: Conducting Field Surveys and Integrating Field Data

As noted in Sect. 4.1, geotagging can be more valuable than a “mere” book-keeping exercise. Hence, this section summarizes points that must be kept in mind while conducting field surveys to maximize the benefits for subsequent spatial analysis.

Objects that can be identified on georeferenced satellite images are effectively geotagged already. When objects that cannot be identified on such images (e.g. because they are too small or are covered by vegetation) are identified at the site, their location can be recorded using a handheld GPS or any other satellite navigation system that provides geospatial positioning anywhere in the world. Modern smartphones have inbuilt receivers to determine latitude, longitude, and altitude to within a few metres of accuracy. (High-precision devices can even achieve precision in centimetres.)

Many archaeological sites are in remote areas, and they are often overgrown with scrub or buried in accumulated dust or silt. Navigating to such sites without familiar reference points (buildings, intersections, etc.) is often tricky, and even available reference points can be camouflaged. Hence, it is advisable to preload points of potential interest at the site (identified ahead of time using georeferenced satellite images) and use the real-time moving maps provided by handheld devices to navigate to these points. It is also useful to log the track, not only to help retracing steps (and revisiting the site later), but because such a log imposes a natural sequence for the order in which features were geotagged. Such a log (which includes timestamps) can greatly simplify the task of integrating data gathered on multiple devices such as cameras.

This usage of satellite navigation (sat nav) played an important role in our geospatial and ground-based archaeological survey at Talakadu, situated at a hairpin bend on the River Kaveri in the Indian state of Karnataka. Over the last few centuries, annual monsoon winds have carried fine sand from the point-bar deposit towards the old settlement, burying it to depths of 2–30 feet. Prior excavations by the Directorate of Archaeology and Museums of Karnataka and the University of Mysore in 1992–1993 had explored less than 1% of the sand-covered area.

There are five well-known temples in this area—Vaidyeshvara, Kirtinarayana, Maralesvara, Pataleshvara, and Chaudesvari—which were constructed between the tenth and fourteenth centuries CE. The first three temples can be readily identified in optical satellite images because they are exposed above the surface, but the latter two temples are hidden under trees. We geotagged the exact locations of the hidden temples and of all the excavation sites. By correlating information from the excavation of specific trenches with larger landscape features identified by analysing the site's spatial context using satellite imagery, we could contextualize information from multiple sources and scales and interpret appropriately (Rajani et al. 2012). This site is further discussed in Sect. 5.4.

4.5 3D Landscape Visualization

Creating a virtual reality 3D landscape visualization can enable the analyst to simulate visualization of the terrain as it would have appeared in the past. This capability of GIS is best demonstrated in the visualization of the Seven Pagodas of Mahabalipuram in Sect. 5.5 to propose a novel hypothesis for the name of this site. Another application is to create "then vs. now" scenes of the same landscape viewed from the same location. Figure 4.11 shows such a simulation of the hill fort of Chitradurga in the Indian state of Karnataka. Perspective views of the hill fort from old paintings (Fotheringham 1801) were simulated using DEM, which conspicuously displays the changes in land use in the vicinity of the hill fort (Nalini and Rajani 2012).

Fig. 4.11 **a** Old painting "View of Chiteldroog from the east, 1801" (© British Library Board; WD581, 581); **b** a simulation of the view depicted in the old painting using DEM and Google Earth Image

This technique also provided unique inputs for studying the history of the landscape of the Lalbagh Botanical Garden in Bangalore. In 1760, Hyder Ali established a garden, which was later expanded by his son Tipu Sultan. After Tipu was defeated and killed in 1799, the stewardship of Lalbagh passed through several British Superintendents until Indian independence. The popular assumption was that gardens laid out by Hyder and Tipu are contained within its present boundaries (Fig. 4.12b). Two maps: (i) *Plan of the fortress of Bangalore*, 1791 and (ii) *Plan showing the position of the British Troops round the Pettah*, March 1791, together show five distinct plots (Fig. 4.12a). These maps have other features (fort, roads, and bunds) which aided in georeferencing them to identify the corresponding area of the plots on satellite images. Shortly after the capture of Bangalore by Cornwallis' Army in 1791, a number of official and amateur British army draughtsmen, surveyors, and artists, painted scenes showing what they called the "Cypress Garden": (1) *East view of Bangalore, with a small shrine and a dismounted horseman in the foreground, and cattle grazing beyond* (1791) by Robert Hyde Colebrooke (Fig. 4.12d) (2) *Southerly view of Bangalore* (1792) Claude Martin, (3) *East View of Bangalore with the Cypress Garden* (1792) by Robert Home (Fig. 4.12e), and (4) *East view of Bangalore, with the cypress garden* (1792), from a pagoda by James Hunter. These four paintings show the Bangalore *Kote* (fort) structure with bastions at the far end, and rectangular patches of greenery with rows of cypress trees in the foreground (although from these paintings it is difficult to discern the exact number of plots). The first among these paintings show the Kempegowda watchtower, and the third shows a temple (both their locations are known) in the extreme foreground, which suggests the artist was close to these structures and illustrated the view visible from this point. Using these locations as references, we created virtual 3D views of the landscape similar to that depicted in the paintings (Fig. 4.12a, c). This further supported our identification of the former plots. Using this kind of geospatial analysis together with Buchanan's account stating that Hyder's gardens were watered from a nearby tank while Tipu's gardens used mechanized water transport (Buchanan 1807), we could

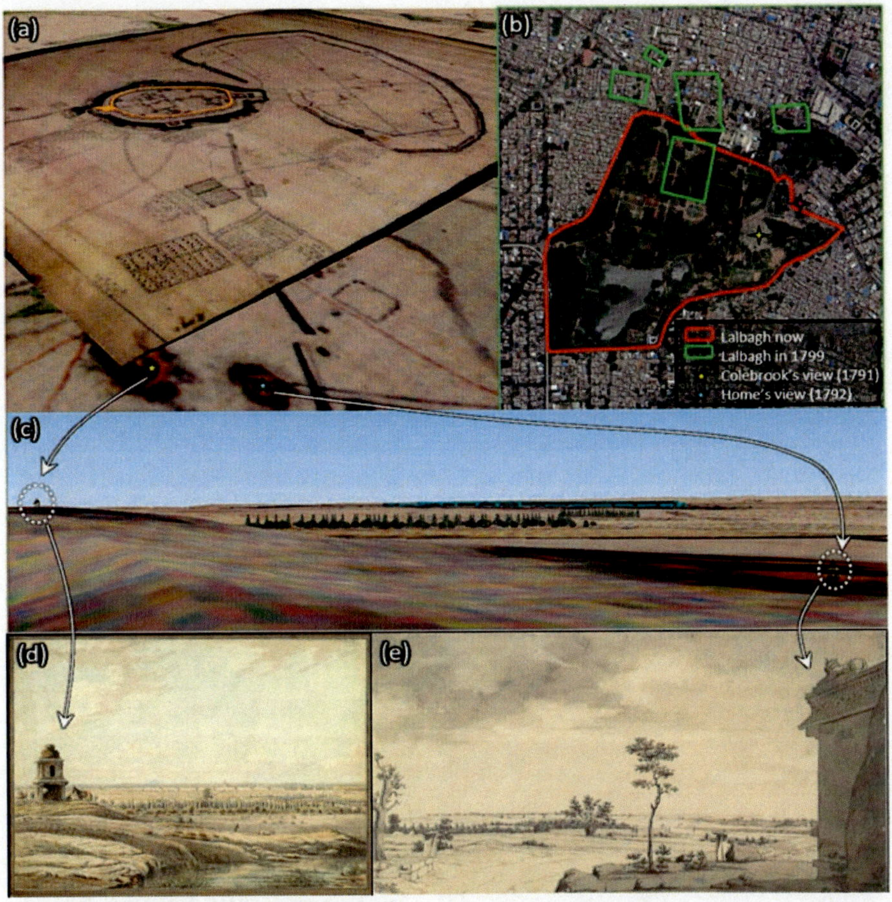

Fig. 4.12 **a** 3D perspective showing the maps of (1791) and the location from where painting by Colebrooke (**d**) (© British Library Board; WD4461, 4461) and Home (**e**) (© British Library Board; WD3775(5), 5) were made; **b** Google Earth image showing the original and present layout of Lalbagh; **c** a simulated view that is similar to view in paintings made by Colebrook and Home

speculate which of the plots were laid out by Hyder and which by Tipu based on their proximity and slope from nearby tanks. Our study has analysed old maps, old paintings, satellite images, and simulated views similar to those recorded in paintings using 3D virtual GIS and found that Hyder and Tipu's gardens comprised of five distinct plots and only one of these garden plots overlap with the modern Lalbagh (Iyer et al. 2012).

4.6 Other Kinds of Spatial Analysis and Modelling

GIS facilitates many kinds of spatial analysis, and we present a few illustrative examples here. For a comprehensive review, we suggest Verhagen (2018).

GIS can extract 2D and 3D surface profiles from the DEM layer. As an example, Fig. 4.13 shows such profiles for Talakadu (discussed in more detail in Sect. 5.4).

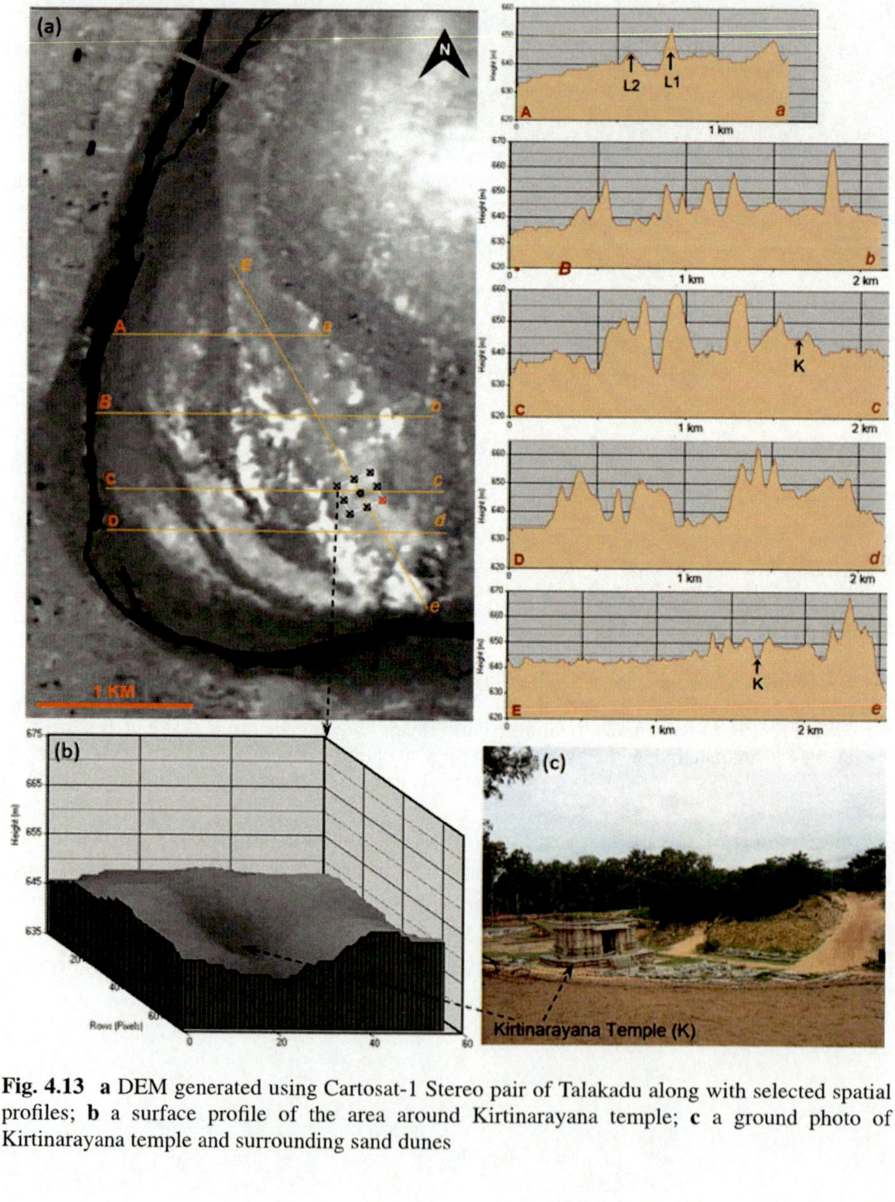

Fig. 4.13 **a** DEM generated using Cartosat-1 Stereo pair of Talakadu along with selected spatial profiles; **b** a surface profile of the area around Kirtinarayana temple; **c** a ground photo of Kirtinarayana temple and surrounding sand dunes

Here, the analyst has drawn 5 transect lines A to E (Fig. 4.13a) across the landscape. GIS can create graphs to visualize the undulations along these lines as pixel frequencies. By selecting an area, the software can similarly create a 3D profile (see Fig. 4.13b). At a site such as Talakadu, which has significant sand accumulation, such an analysis can estimate the volume of sand that must be cleared prior to excavation. Further, one can exaggerate the display of differences in heights represented by the DEM so that the analyst can recognize patterns in undulations more easily.

Using input from a DEM, an analyst can mark a specific vantage point and GIS can generate a viewshed map for a given radius. This creates a layer which indicates which points are visible from the vantage point or occluded due to elevations along the line of sight to the vantage point. This tool is typically used in military and telecommunication applications to identify locations for specific facilities, but it has interesting applications for managing archaeological sites as well. For example, by generating visibility maps for a 350 m radius around five monuments at Badami (Rajani et al. 2009) and six monuments within the Chitradurga fort (Nalini and Rajani 2012), we identified areas that could be selected for developmental activities while preserving the aesthetic views from these monuments, which we recognize as part of their heritage value.

A final interesting example is by Gillespie et al. (2016), who have developed a predictive model using eight environmental parameters to identify 121 possible locations in the Indian subcontinent where inscriptions of the Ashokan era may lie.

References

Albernaz JT, Attayde JD, Seixas YLFD (1630) Taboas Geraes De Toda a Navegação. Control number: 78653638, Library of Congress Geography and Map Division Washington, D.C. 20540-4650 USA. Digital id: http://hdl.loc.gov/loc.gmd/g3200m.gct00052. Accessed 16 Apr 2020

Andrews JH (2009) Maps in those days: cartographic methods before 1850. Four Courts Press, Portland, p 138

Asia Maior (Firm). & Boink G. G. J. & Bos, Jeroen. & Brommer, Bea. & Brink, Paul van den. & Bruijn, Mirjam de. & De Roever, Arend. & Ferwerda, Hans. & Graaff, Isaac de. & Hattingh, Leon. & Knaap, G. J. & Kruijtzer, Gijs. & Lunsingh Scheurleer, Pauline. & Moerman, Jacques. & Nelemans, Bert. & Ormeling, Ferjan. & Robson, S. & Robson-MacKillop, R. & Schilder, Günter. & Sleigh, Dan. & Stroo, Gurt. & Van Diessen, Rob. & Vorstenbosch, Elsbeth. & Waverley, Glen. & Zielstra, Hein. & Nationaal Archief (Netherlands). & Koninklijk Nederlands Aardrijkskundig Genootscha. & Rijksuniversiteit te Utrecht. Faculteit Geowetenschappen (2006) Grote atlas van de Verenigde Oost-Indische Compagnie Comprehensive atlas of the Dutch United East India Company. (vol 1. Atlas Isaak de Graaf/Atlas Amsterdam / [explanatory text] Günter Schilder, Jacques Moerman, Ferjan Ormeling, Paul van den Brink, Hans Ferwerda., p 135). https://catalogue.nla.gov.au/Record/3802570

Berggren JL, Jones A (2000) Ptolemy's geography: an annotated translation of the theoretical chapters. Princeton University Press, Princeton

Beveridge H (1900) The fort of Masulipatam in 1759. A comprehensive history of India, civil, military, and social, from the first landing of the English to the suppression of the Sepoy revolt; including an outline of the early history of Hindoostan, London Blackie:614. https://archive.org/stream/comprehensivehis01beve/comprehensivehis01beve#page/614/mode/1up. Accessed 12 Apr 2020

Bocarro A (1635) Livro das plantas de todas as fortalezas, cidades e povoaçoens do Estado da India Oriental. Retrieved from the Biblioteca Nacional De Portugal. http://purl.pt/27184/3/#/279. Accessed 16 Apr 2020

Brown RM (2005) Paṭnā's Golghar and the transformations of colonial discourse. Arch Asian Art 55(1):53–63

Buchanan F (1807) A journey from Madras through the countries of Mysore, Canara, and Malabar. T. Cadell and W. Davies, London 1, p 46

Buchanan F (1812) Plan of the city of Patna' no.1 in 172 MSS. Eur. D. 97 34 X 20 cm. Foll. 36. Vol III-Maps and Plans

Colebrooke RH (1791) East view of Bangalore, with a small shrine and a dismounted horseman in the foreground, and cattle grazing beyond. http://www.bl.uk/onlinegallery/onlineex/apac/other/019wdz000004461u00000000.html. Accessed 16 Apr 2020

Cunningham A (1871) Four reports made during the years 1862–63–64–65. Vol 1. Printed at the Government Central Press, Simla. https://archive.org/details/report00cunngoog. Accessed 16 Apr 2020

Cunningham A (1880) Reports of tours in the gangetic provinces from badaon to Bihar in 1875–76 & 1877–78, vol 11. Printed at the Government Central Press, Simla, p 156

Das S, Rajani MB (2020) Palimbothra to Patna: reconstructing the ancient landscape using geospatial analysis. In: Rao N (ed) Facets of indian culture. Aditya Prakashan, New Delhi And S R Rao Memorial Foundation For Indian Archaeology, Art And Culture

Das S, Kumar P, Rajani MB, Chopra S (2019) Radiocarbon dating of historical bricks: exploring the unprotected archaeological mounds in the environs of excavated site of Nalanda. Proc Indian Nat Sci Acad PINSA (special section) 85(3) Sept:619-628. https://doi.org/10.16943/ptinsa/2019/49649

Fotheringham RH (1801) View of Chittel Droog from the East. https://www.bl.uk/onlinegallery/onlineex/apac/other/019wdz000000581u00000000.html. Accessed 16 Apr 2020

Gillespie TW, Smith ML, Barron S, Kalra K, Rovzar C (2016) Predictive modelling for archaeological sites: Ashokan edicts from the Indian subcontinent. Curr Sci 110(10)25 May:1916–1921

Gupta E, Rajani MB (2020a) Historical coastal maps: importance and challenges in their use in studying coastal geomorphology. J Coast Conserv 24(24). https://doi.org/10.1007/s11852-020-00739-7

Gupta E, Rajani MB (2020b) Geospatial analysis of historical cartographic data of Kollam Fort. J Indian Soc Remote Sens. https://doi.org/10.1007/s12524-020-01181-w

Harvey PDA (1987) Local and regional cartography in medieval Europe. In: Harley JB, Woodward D (eds) Cartography in prehistoric, ancient, and medieval Europe and the Mediterranean. The History of Cartography 1:495

Hesse R (2016) Talk: Historical maps and ALS visualisations. Cultural Heritage and New Technologies, 16–18 November, Vienna. https://www.academia.edu/29998719/Historical_maps_and_ALS_visualisations. Accessed 24 Jan 2017

Home R (1792) East view of Bangalore with the cypress garden, 5. http://www.bl.uk/onlinegallery/onlineex/apac/other/019wdz000003775u00005000.html. Accessed 16 Apr 2020

Hunter J (1792) East view of Bangalore, with the cypress garden, from a pagoda. http://www.bl.uk/onlinegallery/onlineex/apac/other/019xzz000007683u00008000.html. Accessed 16 Apr 2020

Innes CA (1997) Malabar Gazetteer—vol 1 and 2, Evans FB, I.C.S (1908) (eds) S. Hemachandran (1997), Kerala Gazetteers Department, rpn. 1997, p. 535

Iyer M, Nagendra H, Rajani MB (2012) Using satellite imagery and historical maps to investigate the original contours of Lalbagh Botanical Garden. Curr Sci 102(3) 10 Feb:507–509

Jackson VH (ed) (1925) Journal of Francis Buchanan: kept during the survey of the districts of Patna and Gaya in 1811–1812. Patna:179. https://archive.org/details/in.ernet.dli.2015.530981/page/n5/mode/2up. Accessed 12 May 2020

Jennings W (1755a) A plan of Madura, with the adjacent places round, t. BL: Maps.K.Top.115.87. https://imagesonline.bl.uk/en/asset/show_zoom_window_popup_img.html?asset=18294. Accessed 15 Apr 2020

Jennings W (1755b) A plan of Madura. BL:Maps.K.Top.115.88. https://imagesonline.bl.uk/en/asset/show_zoom_window_popup_img.html?asset=18295. Accessed 15 Apr 2020

Keay J (2000) The great arc. Harper Collins, London

Kirby RS (1990) Engineering in history. McGraw-Hill, New York, p 131. https://archive.org/details/B-001-000-180/page/n139

Longe FB (1906) General report on the operations of the survey of India during 1904–1905. Office of the Superintendent of Government Printing, Calcutta

Malik JN, Gadhavi MS, Satuluri S, Kumar S, Sahoo S, Vikramam B (2017) Unravelling the hidden truth from Vigukot in the Great Rann of Kachchh, western India by surface and sub-surface mapping. Curr Sci 113(10) 25 Nov:1906–1917

Martin C (1792) Southerly view of Bangalore. http://www.bl.uk/onlinegallery/onlineex/apac/other/019pzz000000256u00000000.html. Accessed 16 Apr 2020

Mitter P (1977) Much maligned monsters. Clarendon Press, Oxford

Nalini NS, Rajani MB (2012) Stone fortress of Chitledroog: visualizing old landscape of Chitradurga by integrating spatial information from multiple sources. Curr Sci 103(4):381–387 (25 Aug)

Opreanu CH, Lăzărescu V (2015) A Roman Frontier marketplace at Porolissum in the light of numismatic evidence. Centrul de Cultură și Artă al Jude ului Sălaj. https://www.researchgate.net/publication/309576385_OPREANU_LAZARESCU_2015/stats

Plan de Pondicherry En 1741. https://artsandculture.google.com/asset/puducherry-union-territory-of-puducherry-anonymous/1AF5Zhx92V6CCQ. Accessed 15 Apr 2020

Plan of the fortress of Bangalore (1791) MAPS illustrative of Lord Cornwallis's campaigns in India, 1778–1792. https://www.alamy.com/plan-of-the-fortress-of-bangalore-1791-maps-illustrative-of-lord-cornwalliss-campaigns-in-india-1778-1792-1778-1792-source-add18109c-image227213898.html (British Library Add.18109C). Accessed 16 Apr 2020

Plan showing the position of the British troops round the Pettah, March (1791) MAPS illustrative of Lord Cornwallis's campaigns in India, 1778–1792. https://www.alamy.com/plan-showing-the-position-of-the-british-troops-round-the-pettah-march-1791-maps-illustrative-of-lord-cornwalliss-campaigns-in-india-1778-1792-1778-1792-source-add18109d-image226873913.html (British Library Add.18109D). Accessed 16 Apr 2020

Rajani MB (2016) The expanse of archaeological remains at Nalanda: a study using Remote Sensing and GIS. Archives of Asian Art, Duke University Press 66(1) Spring:1–23

Rajani MB and Rajawat AS (2011) Potential of satellite based sensors for studying distribution of archaeological sites along palaeo channels: Harappan sites a case study. J Archaeolog Sci Elsevier 38(9) Sept:2010–2016 (https://doi.org/10.1016/j.jas.2010.08.008)

Rajani MB, Kasturirangan K (2013) Sea level changes and its impact on coastal archaeological monuments: seven pagodas of Mahabalipuram, a case study. J Indian Soc Remote Sens 41, 461–468. https://doi.org/10.1007/s12524-012-0210-y

Rajani MB, Patra SK, Verma M (2009) Space observation for generating 3D perspective views and its implication to the study of the archaeological site of Badami in India. J Cult Heritage 10(1):e20-e26. https://doi.org/10.1016/j.culher.2009.08.003 (Dec)

Rajani MB, Rajawat AS, Murthy MSK, Kamini J, Rao S (2012) Demonstration of the synergy between multi-sensor satellite data, GIS and ground truth to explore the archaeological site in Talakadu region in South India. J Geomatics Indian Soc Geomatics 6(1) Apr:37–41

Smith R (1814–15) The granary (golghar) at Bankipur, near Patna (Bihar) seen from the river; European officials' houses nearby http://www.bl.uk/onlinegallery/onlineex/apac/other/019wdz000002090u00000000.html. Accessed 16 Apr 2020

Stuart ER (1918) Map reading and topographical sketching. McGraw-Hill Book Company, New York

Suganya K, Rajani MB (2018) Underground water supply system in the late nineteenth and early twentieth century Bangalore, Water Hist Springer 10(4) Dec:291–311. https://doi.org/10.1007/s12685-018-0223-8

Suganya K, Rajani MB (2020) Riverfront gardens and city walls of Mughal Agra: a study of their locations, extent and subsequent transformations using remote sensing and GIS. South Asian Studies, Routledge. doi:10.1080/02666030.2020.1721119

Taarant/Tarand HG, Duyrendaal LN (1687) Representation of Fort Coylan "D'Grond Teekening van de Fortresse Coylan". Nathional Archive, Netherlands. NL-HaNA_4.VEL_912 (old number: VEL0912). https://commons.wikimedia.org/wiki/File:AMH-2604-NA_Representation_of_Fort_Coylan.jpg. Accessed 16 Apr 2020

Thuillier HL, Symth R (1875) Manual of surveying for India detailing the mode of operation on the trigonometrical, topographical and revenue surveys of India. Thacker Spinck and Co., Calcutta

Verhagen JWHP (2018) Spatial analysis in archaeology: moving into new territories. Bubenzer O, Siart C, Forbriger M (eds) Digital geoarchaeology. New techniques for interdisciplinary human-environmental research, pp 11–25. https://doi.org/10.1007/978-3-319-25316-9_2

Wheeler EO (1955) The survey of India during the war and early reconstruction 1939–1946. Survey of India, Dehradun

Wheeler GM (1885) Third international geographical congress and exhibition at Venice, Italy 1881. Resolution of Congress, Washington

Williams JED (1994) From sails to satellite: the origin and development of navigational science, OUP, New York (1992)1994

Chapter 5
Case Studies

Preamble
In our experience, there is no fixed or formulaic approach to geospatial analysis of archaeological sites because the process of analysis is sensitive to variations between sites. These variations can be grouped into five principal categories: (1) variations in the features of interest (these will vary across sites in terms of their scale, their architectural styles, layout patterns), (2) variations in the natural setting and subsequent natural changes (sites vary in terms of their terrain—flat plains, hills, coasts—as well as their exposure to natural processes such as weathering and floods), (3) anthropogenic changes (sites will vary in terms of subsequent human activities and their impacts), (4) satellite imagery (data from satellites may be plentiful for some sites but limited for others), and (5) historical spatial records (the quality and quantity of such records may similarly vary across sites). The purpose of this chapter is to illustrate how the process of analysis responds to these variations. Thus, we have carefully chosen five sites so that there are significant variations across these sites in each of these five categories. The five sites are Nalanda, Agra, Srirangapatna (Lalbagh Palace), Talakadu, and Mahabalipuram, and before delving into the case studies, we briefly discuss how these sites differ within each category.

Features of interest. The features of interest to the analyst will clearly vary from site to site (and even from study to study for the same site). For our studies at Agra and Srirangapatna, the features of interest are gardens which are in the Persian *Charbagh* (four-part) style. In contrast, we examine large Buddhist stupas and smaller monasteries at Nalanda, and free-standing temples at Talakadu and Mahabalipuram. The sand accumulation at Talakadu makes it difficult to identify smaller structures (although we do see larger-scale features such as palaeochannels), whereas smaller structures can be seen quite easily at Mahabalipuram. Further, our attention at Mahabalipuram is on the locations of temples rather than on the temples themselves.

M. B. Rajani, *Patterns in Past Settlements: Geospatial Analysis of Imprints of Cultural Heritage on Landscapes*, Springer Remote Sensing/Photogrammetry,
https://doi.org/10.1007/978-981-15-7466-5_5

Natural setting and natural changes. Three of the sites we study are located near rivers: Talakadu, Agra, and Srirangapatna. At Talakadu, the river has moved considerably and there is substantial sand accumulation whereas very little interim movement of the river or sand accumulation is observed at Agra and Srirangapatna (the latter site is on a rocky river island). Nalanda is located on the plains well away from the nearest river, whereas Mahabalipuram is located on the coast and is subject to coastal processes with very different dynamics (see Sect. 2.3.7).

Anthropogenic changes. The spatial contexts at Agra, Mahabalipuram, Srirangapatna, and Talakadu have all been heavily disturbed. At the first two sites, these changes happened due to urbanization (e.g. the laying of railway lines and roads at Agra and the construction of a canal at Mahabalipuram). At the latter two sites, the changes were caused by deliberate actions by the British over comparatively shorter timescales (the destruction of Tipu Sultan's palace at Srirangapatna following his death (Moienuddin 2000), and the planting of vegetation at Talakadu to arrest sand accumulation—this vegetation now thickly covers much of the site (Devaraj et al. 1996)).

Satellite imagery. With a wealth of satellite imagery now available, the primary variability in this category is often dictated by the funds available to procure images necessary for the analysis. While we have been fortunate to have adequate funding, our studies at Talakadu and Mahabalipuram were conducted about a decade earlier than at the remaining sites. As the quality and quantity of satellite images have improved in the interim, our case studies vary in this category as well.

Historical spatial records. Sites such as Nalanda and Agra are well studied by scholars such as Frederick Asher and Ebba Koch, respectively, and there is a wealth of historical spatial records (e.g. old maps and textual records) for both sites. In contrast, there is comparatively little academic scholarship for the features of interest at Srirangapatna, Talakadu, and Mahabalipuram. Further, the quantity and quality of historical spatial records for these sites vary considerably from Srirangapatna (a good record consisting of old maps, paintings, and textual records) to Mahabalipuram (a limited record consisting of portolan charts and legends) to Talakadu (a very scant record containing only one earlier excavation report prior to our work). Historical spatial records are particularly valuable for sites that have witnessed substantial anthropogenic change. Without maps for Bangalore (Fig. 3.7), or Kollam (Fig. 4.10), or Agra (Fig. 5.3), it would have been extremely difficult to identify the original settlements in these urbanized sites. Even maps made centuries after the original settlement, such as Colin Mackenzie's 1806 map of the twelfth-century site Halebidu, are valuable (Das and Rajani 2020) because they predate industrialization (which hastened the pace of anthropogenic change). Urbanization was slow to affect much of India until well into the twentieth century, so Corona satellite images from the 1960s and 1970s are often invaluable (see Sect. 3.2.3 and Fig. 3.12).

5.1 Nalanda

Nalanda was a Buddhist monastery of considerable repute, and accounts of visitors from China (dating from the seventh century CE) suggest that it was a large, thriving establishment whose physical dimensions were immense. This institution seems to have remained in existence from the fourth/fifth century CE to at least until the end of the twelfth century CE, but we do not have a continuous record for its activities (Rajani 2016). The Tibetan monk Dharmasvamin records some lingering activity in the monastery during his visit in 1234–36 CE, but the subsequent historical documentation about Nalanda is sparse until the nineteenth century CE.

In 1812, Francis Buchanan surveyed the region that would later be known as Nalanda. The first translations of the accounts of the Chinese travellers Faxian (337–422 AD) and Xuanzang (596–664 AD) into English (Laidley 1848; Beal 1884) provided impetus to the growing interest in the discovery of Indian antiquity among British explorers, and aided by these translations (Laidley 1848), Markham Kittoe visited the site in 1847–48 and identified Baragaon as Faxian's "Na Lo". Subsequently, Alexander Cunningham identified these remains as the ruins of the famous Nalanda that Xuanzang visited. Alexander Meyrick Broadley visited Nalanda in 1871, and his records, together with the records of Buchanan and Cunningham, provided the basis for Archaeological Survey of India's phased excavations beginning in 1863 and most recently in 1983 (Rajani 2016).

The excavated remains comprise sixteen large structures: a row of four temples or *chaitya* on the west (T3, T12, T13, and T14), a row of eight west-facing monasteries or *vihara* (numbered 1, 4, 6, 7, 8, 9, 10, and 11) parallel to the temples, and two smaller north-facing monasteries (1A and 1B), and to the east of monastery 7 are T2 and the Sarai Temple (Fig. 5.1 inset). If Nalanda sustained anywhere near the 10,000 residents recorded by Xuanzang (Beal 1914) or even the 3000 residents recorded by Yijing (Takakusa 1896) solely within the currently excavated extent, the monasteries would have been multi-storeyed premodern skyscrapers which is unlikely (Asher 2018). Hence, many investigators hypothesized that the site was much larger than the currently exposed archaeological remains (Stewart 1989). Therefore, our study aimed at trying to identify the site's extent and to determine whether the remains of any other temples or monastic structures survive.

5.1.1 *Historical Spatial Records and Preliminary Analysis*

While the earliest spatial records are textual, the illustrations made by Buchanan, Cunningham, and Broadley are the earliest available graphical records of the expanse of archaeological ruins in Nalanda's spatial context. Subsequently, the ASI has prepared an architectural drawing of the excavated structures "Nalanda: Excavated remains" published by Misra (1998).

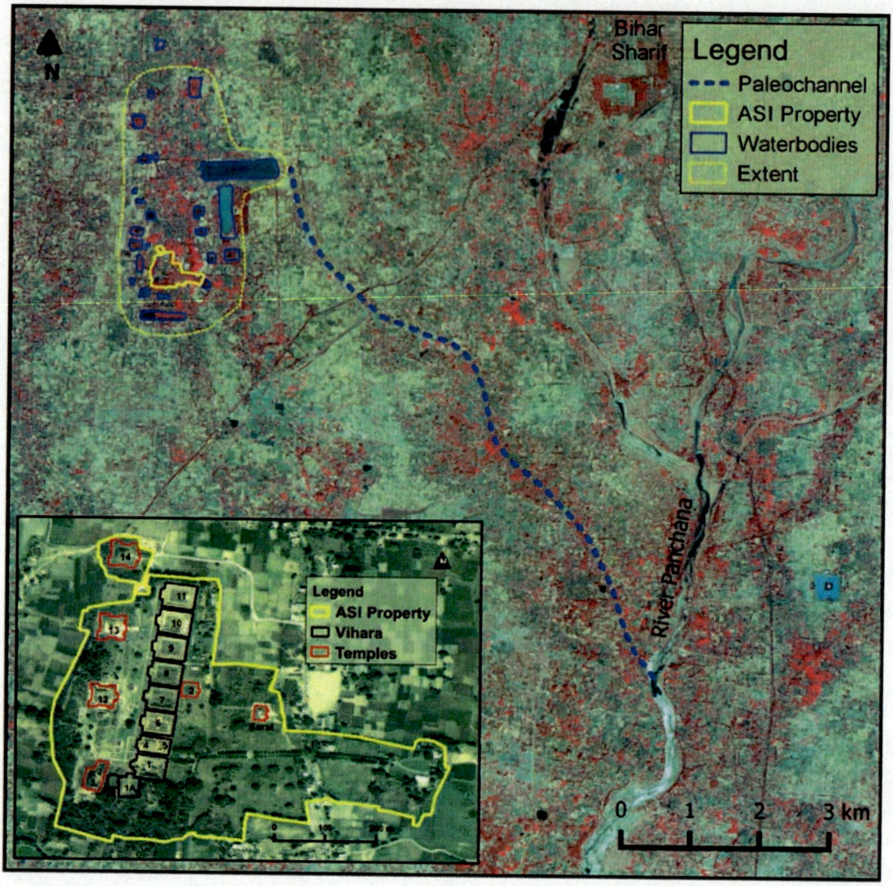

Fig. 5.1 Excavated site of Nalanda (inset) in its spatial context amidst a cluster of water bodies and a palaeochannel

We began our analysis with Google Earth by identifying all excavated structures (temples and monasteries) and marking these as polygons. This gave us a sense of the spatial pattern and distribution of these known features. As we zoomed out to see the site's spatial context, we identified a cluster of water bodies (and former water bodies) whose geometrical shapes (square or rectangular) and layouts (with sides roughly parallel to the cardinal directions) strongly suggested that these were man-made tanks. Given their proximity to Nalanda, we hypothesized that these tanks suggest a rough extent of the establishment.

To strengthen our belief in this hypothesis, we examined the wider region of Southern Bihar and found that such a pattern of tanks was unique to Nalanda. For this part of the analysis, we found that coarse-resolution images were adequate to identify these tanks and were able to procure such images from Landsat and IRS 1C/1D for multiple dates (to investigate seasonal variations). Visible band Landsat images are available online from several portals, including Google Earth and LandsatLook[1] Viewer. By examining the larger area, we were able to identify that at least one tank around Nalanda received water from the River Panchana via a palaeochannel (Fig. 5.1). Further details are presented in Rajani (2016) and Rajani and Das (2018).

Having identified a plausible region of interest (i.e. the area framed by these tanks), we started to procure higher-resolution multispectral images (we used images from the LISS-4 sensor aboard Resourcesat-1 and 2) to identify buried structures via cropmarks. We also generated a DEM (using a stereo images from Cartosat-1) and studied it using anaglyph photogrammetry to identify mounds and anthropogenic shapes of potential archaeological interest.

5.1.2 Generating Novel Hypotheses

The spatial data from available historical records and the above analysis of satellite imagery was integrated using GIS to generate several novel hypotheses. The synoptic view of Nalanda's spatial context (Fig. 5.2) revealed several interesting patterns and hitherto unknown features. First, we observed that the temples T3, T12, T13, and T14 were located at regular intervals along a straight line. Second, we noted that cropmark features F1, F2, and F4 as well as the unexcavated mound F3 also lay at similar intervals along this line. We therefore hypothesized that there were additional temples at these locations in Nalanda. Third, we identified a mound F9 to the north (Fig. 3.16) with an anthropogenic shape: a squarish structure with projections (or perhaps independent structures) at each of its four corners. This mound covers an area of 450 m $\times$ 400 m, and its shape and orientation are very similar to two well-preserved Pala monasteries at Vikramasila (in Bihar) and Somapura (in Bangladesh). We therefore hypothesized that this is the location of another Pala monastery.

5.1.3 Fieldwork

The objective of our field visit was to identify ground-based evidence for or against our hypotheses for features F1, F2, F3, F4, F5, and F9. Prior to visiting the site, we fed the locations of these features of interest to our handheld GPS devices (Sect. 4.4), as well as other reference locations to aid navigation.

F3 is an unexcavated brick mound whose existence has been recorded before, but we could confirm that a structure still existed at the time of our visit. The cropmarks F1 and F2 lie in agricultural fields and are only visible on satellite images of certain seasons. We were unable to identify any evidence for structures at these locations. While the absence of evidence weakens our belief in the hypothesis, it does not refute it. It is possible that future excavations or GPR studies can identify evidence

[1] https://landsatlook.usgs.gov/. Accessed on 13 May 2020.

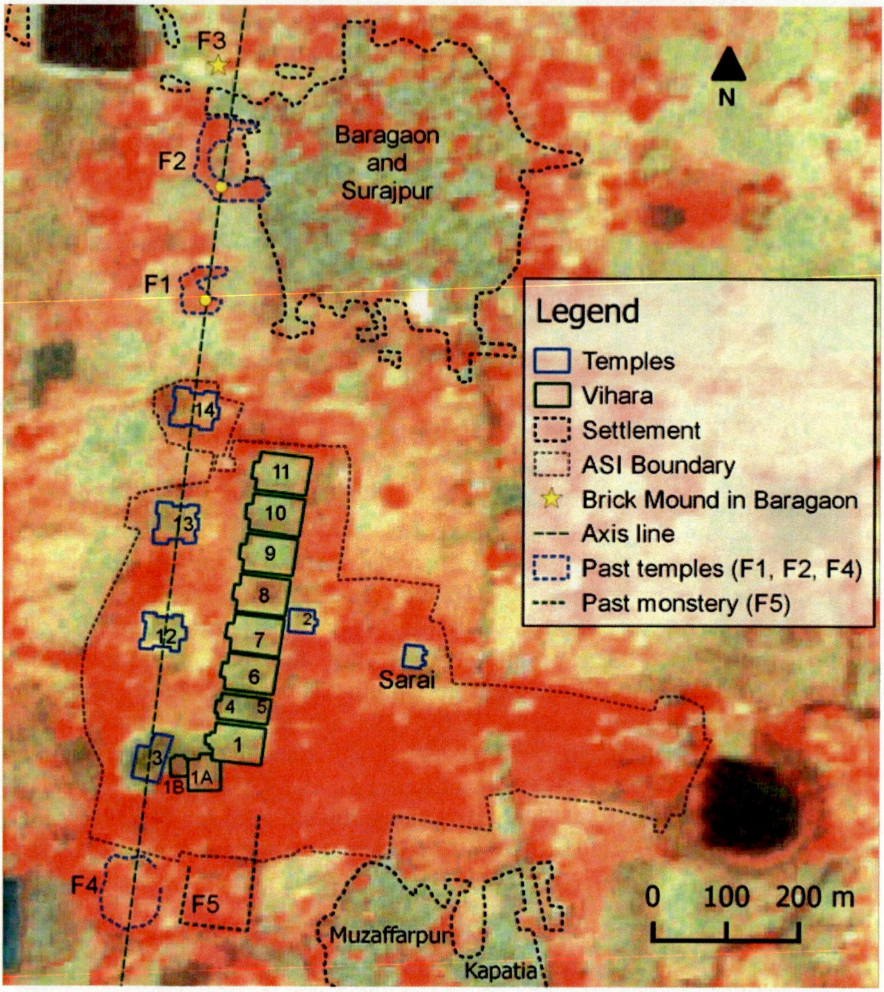

Fig. 5.2 Unexcavated mounds (past temples and monasteries) in the immediate vicinity of Nalanda

for buried structures at these locations. In contrast, features F4 and F5 are in the middle of dense vegetation that we did not have the means to penetrate, but we found enough evidence of undulations on the ground to suggest that there are remains of structures buried at these locations.

During our field visit, we also identified three circular mounds (features F6, F7, and F8) which we could not identify during our geospatial analysis because they were obscured by dense canopies of vegetation. After the field visit, we went back to our geospatial analysis to understand these features in the site's spatial context. In fact, we often find that ground visits lead to further interrogation of images, which may cause us to refine our hypotheses or develop fresh ones. Hence, the overall analysis rarely follows a simple linear trajectory.

The northern half of feature F9 lies below a modern settlement (Begampur). We therefore started our investigations along the southern periphery, but we were unable to find any evidence for the hypothesized buried structure. We then spoke to some residents of Begampur, and one of them took us to his plot of agricultural land where some bricks had been unearthed while ploughing the soil. Our GPS location identified this location close to the north-eastern corner of our hypothesized Pala monastery.

5.1.4 *Interpreting Findings in the Historical Context*

To understand the importance of our findings, it was necessary to place them in the context of prior historical scholarship for the site.

Once we had visited the locations of the newly found features, we sought to establish a correspondence between these and earlier reports by Buchanan, Cunningham, and Broadley (Rajani and Das 2018). We found that each of these records gave unique information, since each of them recorded features they were most interested in. For instance, published versions of Buchanan's handwritten diary (Jackson 1922) have omitted his freehand sketch of Nalanda, which we were able to access from his original manuscript (Buchanan 1812). Analysing this sketch added considerable clarity to his text and improved our understanding of the spatial context of the time. Further, in all past academic literature, the mound annotated as Y in Cunningham's map (Cunningham 1871) is interpreted as Temple 2 as both lie east of the row of monasteries (Fig. 5.1 inset). However, our spatial analysis indicates the mound Y was in fact the mound that contained Sarai Temple before it was excavated. Hence, it is necessary to revise all past excavation reports of both these sites (T2 and Sarai) with this new information.

5.2 Agra

The Taj Mahal was just one (albeit the most celebrated) monument among many glorious architectural specimens in early eighteenth century Agra, which included several riverfront gardens. There is a strong demand for land in the vicinity of sites of historical interest, particularly World Heritage Sites, and Agra has witnessed several large-scale development activities since the nineteenth century CE. These developments, which include the laying of railway lines, roads, and urban/industrial sprawl, have largely disregarded the overall cultural landscape. As a result, Agra's protected riverfront monuments—Taj Mahal, Agra Fort, Itimad-ud-Daula's Tomb—survive as disconnected islands, while the few riverfront gardens that remain are rapidly deteriorating into obscurity. Our interest in these gardens was triggered by a map of Shahjahani Agra made in 1720 (Fig. 5.3a), which recorded several such

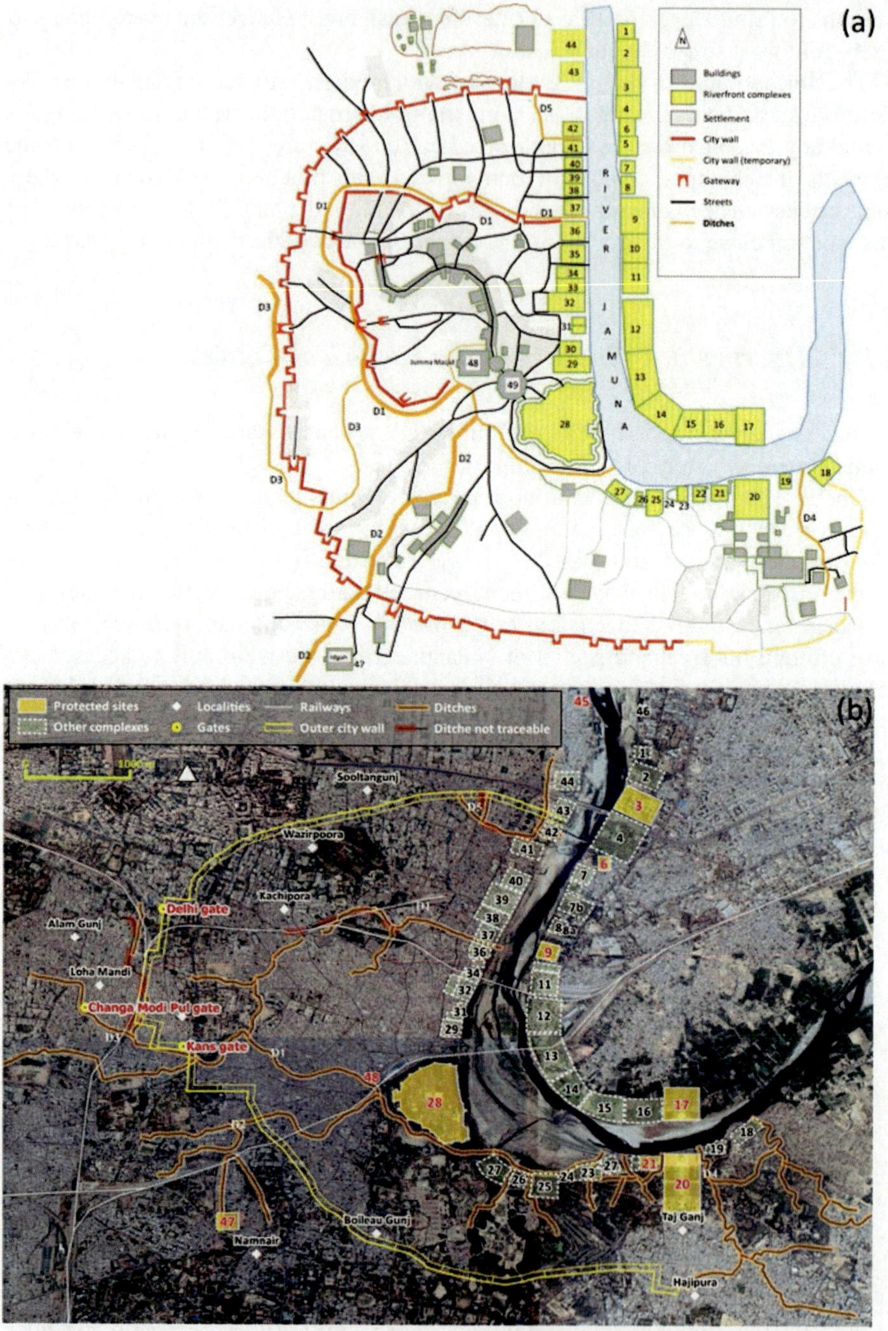

Fig. 5.3 **a** A drawing adapted from published copies of the Jaisingh map of 1720; **b** a Google Earth image showing the locations of key features from the Jaisingh map

gardens. Therefore, our study aimed at identifying the precise locations for these gardens (Suganya and Rajani 2020). For our study, we only had access to published copies of this map. With access to the actual map (or a high-resolution scan), it may be possible to conduct a more detailed geospatial analysis than the one presented here.

5.2.1 Historical Spatial Records

The 1720 map was made for Sawai Raja Jai Singh II of Jaipur (Gole 1989; Koch 2006; Peck 2011), the governor of Agra between 1722 and 1737. This map shows the plan and elevation of buildings as well as their architectural features with annotations in Devanagari script, and it is perhaps the earliest depiction of the larger city (Koch 1986, 1997). We will refer to this as the Jaisingh map, but it is also known as the Jaipur map (Koch 1986, 1997, 2005, 2006, 2008; Koch and Losty 2017). It marks *bagh* (gardens), *rauza* (tomb gardens), *havelis* (mansions), and a *Qila* (the citadel popularly known as the Agra Fort) along the banks of the River Yamuna, as well as two tiers of city walls. Some of the other structures depicted in the Jaisingh map survive today and are easily identifiable: the Idgah [47][2] (south-west of outer city wall), Jama Masjid [48] (north-west to the fort), and the octagonal Tripolia Bazaar [49]. Koch (2006) has identified and numbered forty-four [1-44] of these riverfront complexes and the Jaswanth Singh's Chhatri [45] (which lies beyond the extent of the Jaisingh map), all shown in Fig. 5.3.

Apart from the Jaisingh map, the historical spatial record for Agra includes a scroll painting (1827–35) found in the British library (Koch and Losty 2017) as well as three maps published by Survey of India in 1870, 1944, and 1972. For brevity, we will call these the Scroll[3], the 1870 map, the 1944 map, and the 1972 map, respectively. These, coupled with modern satellite images, provide semi-centennial snapshots of the evolution of Agra's spatial context.

We also have textual records that supplement our understanding of this evolution. Riverfront gardens were introduced by Babur, the first Mughal emperor. By Shahjahan's reign (1628–1658), both banks were fully developed as a stretch demarcated for the imperial family and nobles who served the court (Koch 2006), while the rest of the city lay to the west of the U-shaped meander of the river. The titles of the gardens were based on merit and the ownership changed with time (Koch 2019), and some of the names mentioned in Pelsaert's record (Moreland and Geyl 1925) can be corroborated with those in the Jaisingh map and the Scroll. Although these complexes were thriving when Bernier (1659–63) (Smith 1916) and Tieffenthaler (1744–47) (John John 1997) visited Agra, letters written in 1727 by agents of Raja Jai Singh II suggest that the city and a few of its tombs were being robbed regularly

---

[2]Numbers within square brackets refer to annotations in Fig. 5.3.

[3]The Scroll annotates two sequences of numbers for riverfront façades: 1-32 running north–south on the right and 1-18 running south–north on left banks. We refer to these as R1-R32 and L1-L18, respectively.

(Khan 1993). In 1783, William Hodges described Agra in ruinous condition "as far as the eye can see". He noted that the fort was gradually going into "dissolution" that Dara Shikoh's haveli [29] was in a ruinous state, but the central building of the Taj Mahal was in good form with prayers taking place at its mosque (Hodges 1793). Similarly, Twining visited Agra in 1794 and found the city in ruins apart from the fort, the Taj Mahal and its garden (Twining 1893). The Scroll corroborates these observations by depicting several riverfront gardens in ruinous conditions.

The mid-nineteenth century saw major developmental activities that impacted the spatial context of the riverfront gardens. In 1837, the right bank was cleared of riverfront complexes to construct Strand Road (now Yamuna Kinara Road). The haveli of Asaf Khan and parts of the Jama Masjid were brought down for security reasons during the mutiny of 1857 (Carlleyle 1874), and the haveli of Dara Shikoh was brought down in 1881. The riverfront gardens have continued to decay as the city has continued to develop rapidly.

5.2.2 *Preliminary Geospatial Analysis*

Our first task was to identify the spatial extent represented by the Jaisingh map. First, we identified all protected monuments that also featured in the Jaisingh map, on Google Earth imagery (Fig. 5.3b). Next, we tried to identify the riverfront gardens in these satellite images. While some complexes were discernible, the dense urban landcover made it difficult to identify the boundaries of several gardens, and we were even less successful in identifying indications of the two city walls using these images (Suganya and Rajani 2020).

Therefore, we examined the overall landscape to identify an appropriate strategy. At Agra, the River Yamuna flows from north to south, before taking a U-shaped meander towards the north-east. The region depicted by the Jaisingh map can be split into two regions based on the main type of landcover. Region 1 lies to the west of the U-shaped meander and is a densely urbanized settlement. Region 2 is the inner portion of the U-shaped meander and mainly consists of agricultural and plantation parcels with intermittent settlements. For Region 1, we recognized that the high-resolution natural colour imagery available in Google Earth would be appropriate for identifying spatial patterns in streets and ditches. For Region 2, however, we required additional multispectral imagery. We also obtained Corona imagery in case some features were visible before the pace of urbanization picked up aggressively in the last few decades.

5.2.3 *Detailed Geospatial Analysis*

We used Koch (2006) and Koch and Losty (2017) as base references for the sequence of riverfront complexes and their titles. We first identified the boundaries of eight monuments protected by ASI [3, 6, 9, 17, 20, 21, 28, 45]. These properties have well-defined boundaries that we could identify with great precision (Fig. 5.3b). Next, we explored the complexes whose layouts are still discernible as patterns in vegetation [1, 2, 8, 14-16, 18, 19, L10-1/2, L10], or complexes where the riverfront facade or a tower or a wall provides some reference to trace its boundaries [2, 4, 7, 22, 35, 44, R27, R26]. Finally, we filled the gaps by correlating patterns in the Jaisingh map with the subtle features on satellite images like street patterns in the built-up area [10-13, 29-34, 36, 37, 39-41] or examining the terrain and drainage patterns [23-27, 38, 42, 43].

Further, we tried to identify the locations of the two city walls encircling the citadel as marked in the Jaisingh map. These walls formed two concentric loops, whose ends met the river between specific riverfront complexes (the inner wall is wedged between [36-37] and [28-27], whereas the outer wall begins between [42 and 43] and goes round anticlockwise towards [18]). By first establishing the locations of the riverfront complexes, we were much more successful in tracing the walls using drainage and street patterns. We also used direct or indirect indicators marked in old SOI maps for the locations of *darwaza* (gateways) that provided access to the walled area. Some gateways were marked in the 1870 map (Jamuna, Bheron, Madar, Mania, Nim, and Maithan) but not in the later maps (1944 and 1972). However, their names have been attributed to features that exist to this day at approximately the same locations. For instance, Noori, Bans, Maithan, Neem, Bheron and Jamuna darwaza (gates) have no identifiable physical traces, but their names have survived as Noori Darwaza Road, Bans Gate Area, Maithan Gurudwara, Neem Wali Masjid (mosque), Bhairon Bazaar, and Bhairon Nala (Suganya and Rajani 2020).

5.2.4 *Fieldwork and Interpreting Findings in the Historical Context*

Based on existing records, we have established geospatial positioning for all the 45 riverfront complexes and two tiers of city walls marked in the Jaisingh map (Suganya and Rajani 2020). We find identification of locations and extents by Koch (2006) and Koch and Losty (2017) is entirely consistent with our study for ten of these complexes, but we were able to identify the extent of three complexes more fully and provide locations and approximate extents for a further twenty-seven complexes. Further, we found that the remaining complexes were erroneously labelled, or missing, or misinterpreted in one or more of existing sources. While Koch (2006) has recognized some of these mistakes, our analysis has enabled us to identify further flaws and suggest probable locations and extents for these complexes (Suganya and Rajani 2020). We are yet to conduct focused fieldwork to find evidence for or against our hypothesized locations for the city walls and the few locations that we identified for the first time (Shah Guda's Tomb, Hurkishen temples, Boatman's Shrine, Raj Ghat).

5.3 Lalbagh Palace at Srirangapatna

Srirangapatna is a river island that was the site of four Anglo-Mysore wars between the British and Tipu Sultan. This section will describe only a part of our broader study of Srirangapatna (Gupta et al. 2017), focusing only on the Lalbagh Palace. Unlike the well-known Lalbagh Botanical Garden in Bangalore (originally laid out in the eighteenth century CE by Tipu Sultan's father Hyder Ali and then extended by Tipu Sultan; see Sect. 4.5), its contemporary namesake in Srirangapatna was forgotten (Swamy 2010). The palace had been described as "the handsomest building" of the time by Francis Buchanan (1807). In the nineteenth century CE, it was demolished by the British and its remains became "a mute witness to the acts of vandalism of the British Soldiers" (Moienuddin 2000).

5.3.1 *Historical Spatial Records*

Srirangapatna was of immense military importance to the British, as Tipu Sultan put up tough resistance (see also Sect. 7.2). This site was therefore well documented in the 1780s and 1790s in the form of maps of the fort and its environs, as well as drawings and paintings. These records were probably used for planning the attack in the war in which Tipu Sultan was killed. Unfortunately, very few of these spatial records have details of the extreme eastern corner of the island where the Lalbagh Palace was located. The palace is marked, however, in a nineteenth-century map *Plan of Seringapatam* (Rice 1897).

Some of these British records are now scattered in archives across the world. For instance, we found some interesting material on Srirangapatna from the Macquarie University Library in Australia.[4] While scanning these archives, it was necessary to use the British rendering of the name "Seringapatam". We carefully sieved through all available maps, drawing, and other visual material. The most useful records were the *Plan of Seringapatam* (1897) which marks the "Site of Tipu's Lalbagh Palace" with paths extending in four cardinal directions, an etching entitled *A View of the Island Fort of Seringapatam* (1792)[5] as seen from the north-east direction, with a dense plantation canopy including rows of cypress trees on the extreme left-hand side (which may have lined the radial paths as they did at Lalbagh in Bangalore) (Iyer et al. 2012), and an aquatint entitled *Garden Gate, Laul Baugh, Seringapatam* depicting the entrance gate to the Lalbagh Gardens at Srirangapatna flanked by two dovecotes (Hunter 1792). We note that we only found the latter record after finding evidence for such structures during our fieldwork (Sect. 5.3.3).

[4]https://www.mq.edu.au/macquarie-archive/seringapatam/intro.html. Accessed on 13 May 2020.

[5]http://www.bl.uk/onlinegallery/onlineex/kinggeorge/a/003ktop00000115u06700002.html. Accessed on 17 April 2020.

5.3.2 Geospatial Analysis

From the *Plan of Seringapatam* (1897), we noticed that the palace lay to the east of the complex of the Gumbaz (the mausoleum of Hyder Ali and Tipu Sultan). When viewing this area using Google Earth images, we found that it was largely agricultural, but divided into small parcels. Thus, we recognized that a combination of multispectral imagery to search for cropmarks and high-resolution images to detect field boundaries would be appropriate. We used multispectral images from LISS-4 (5.8 m spatial resolution), but we were unable to see patterns within the garden area. We then acquired images from World View 2, whose pansharpened images have a spatial resolution of 0.5 m. An analysis of these images clearly marks a shape that correlates well with Lalbagh Palace's layout as depicted on the map (Fig. 5.4c). As is typical for the Persian *Charbagh* (four gardens) architecture, it has a central structure with axial paths dividing the surrounding large square garden into four parts. The synoptic satellite view from Google Earth (Fig. 5.4d) also shows a pattern in the agricultural field boundaries: the land within and outside this square layout has been divided into smaller parcels without disturbing the overall boundary, so the layout's original shape is preserved (as discussed in Sect. 2.3.2).

5.3.3 Fieldwork

In preparation for ground truthing, we preloaded handheld GPS devices with the locations we wanted to visit, together with some waypoints to access those locations. These waypoints proved particularly useful during our field visit. Lalbagh Palace was in the middle of a large grid of fields, and we wanted to reach the location "17" (Fig. 5.4c, d). On the satellite image, we could see a road skirting these fields to the north and we first tried this approach. After passing the entrance to the tomb complex and Bailey's memorial (marked as "16" in Fig. 5.4c, d), we saw a weathered dovecote standing to the south of the road amidst fields and bushes. We found the second dovecote in a collapsed state to the north of the road, suggesting that these were originally erected symmetrically on either side of the road. From this road we could not find a path among the fields to reach our target location "17", but fortunately we had planned an alternate route (marked "15" in Fig. 5.4c, d) via the Gumbaz. A ground survey at our target location identified the basement of the palace. Although it was completely overgrown, several architectural ruins were scattered around.

5.3.4 Interpreting Findings in the Historical Context

The pointer provided by the old map of 1897 followed by interpretation of satellite imagery helped us identify the location and the *Charbagh* layout of the garden.

Fig. 5.4 Ground photographs of **a** the collapsed ruins of the northern dovecote and **b** the southern dovecote; **c** a portion of *Plan of Seringapatam* (1897); **d** a Google Earth image showing features of Lalbagh Palace marked in (c); **e** the aquatint *Garden Gate, Laul Baugh, Seringapatam* (1804) by James Hunter (© British Library Board; X768/3(6), 30006)

Ground truthing the location led us to find not only ruins of the palace but also the dovecotes, whose context (flanking the garden entrance) could be corroborated with the painting that illustrates them. Clearance of overgrowth and excavation may reveal the layout and remains of this forgotten palace.

5.4 Talakadu

Talakadu is an archaeological site near Mysore situated on the banks of the River Kaveri. The river meanders in a hairpin bend south-west of the site, and deposits point bar sand on the inner bank. Thus, the archaeological settlement is smothered in wind-blown sand to depths between 2 feet to more than 30 feet, covering an area of 4.5 km^2. In the early nineteenth century CE, trees were planted to arrest the movement of sand.

There are two main theories that explain such large sand accumulation: (1) the annual south-west monsoon wind pattern has swept the fine sand over to the site on the north-eastern side in the last few centuries burying it under several feet of sand (Sivaramkrishna 2005), and (2) faults close to Talakadu suggest that the region is structurally controlled and resulting tectonic activities, though mild, has caused subsidence of ground in the Old Talakadu sinking the area that consists of temples and remains of the archaeological settlement. The subsidence in the ground level has caused accumulation of loads of sand brought by flood waters (Valdiya 2008).

There are five well-known temples associated with this site, constructed between the tenth and the fourteenth centuries CE. The Directorate of Archaeology, Museums and Heritage, of Karnataka together with the University of Mysore conducted excavations at seven locations in Talakadu (numbered TK 1 to TK 7) in 1992–93 (Devaraj et al. 1996). While these excavations advanced understanding of the nature, history, and archaeology of the site, only less than 1% of the sand-covered area was excavated. Our study of this site in 2006–10 overlapped with another excavation in 2006–09 (Murthy et al. 2019). We collaborated with these efforts, and we had two main questions of interest regarding the use of geospatial analysis at this site. First, could such an analysis identify settlements that are now buried under the sand and vegetation canopy? Second, could the analysis suggest potential sites for further excavation?

5.4.1 *Historical Spatial Records*

The only spatial record we could find was the report from the 1992–93 excavation, which was published in 1996. Apart from listing all the findings from specific trenches, there was one hand-drawn map (not to scale), showing the site's spatial context (the river and the sand-covered area).

5.4.2 Geospatial Analysis

When we chose to work on this site, our belief was that since the sand cover had obscured surface telltale signs to a far greater extent than other sites, remote sensing may be capable in identifying new features at this site.

We began by identifying the coordinates of the seven trenches TK 1 to TK 7 excavated in the 1990s and the locations of the known temples so that they could be studied in context with georeferenced satellite imagery (Rajani et al. 2012). We found it easiest to identify these by recording GPS locations during an initial field study, since the features were hard to identify among canopy with the available satellite imagery.

Due to the variety of landcover at this site, we decided that a combination of data from several sensors was necessary for the main geospatial analysis. We chose multispectral imagery to identify patterns of buried settlements below the canopy cover, microwave imagery to penetrate the sand cover, and DEM to see whether 3D visualization could identify patterns in the sand accumulation to shed new light.

Using multispectral imagery, we identified four promising features for further archaeological exploration (Fig. 5.5) (Rajani and Patra 2009). First, cropmarks indicate a linear feature (L1) that is 2.4 km long and 5–10 m wide connecting the Madhava Mantri Dam and the area of Old Talakadu. Considering that the dam was originally built in the fourteenth century CE, we hypothesize that L1 was a canal through which water was supplied from the reservoir to the Old Talakadu area. Second, a rectangular feature (R) measuring about 170 m × 200 m appears as a negative cropmark with a positive cropmark in its centre (distinctly observed in coarse-resolution LISS-3 multispectral images), and we hypothesize that this may have been a reservoir. Third, a patch of vegetation (VP) that is 1.2 km length and 100 m wide lies in a NW-SE direction across Old Talakadu. This patch includes parts of L1 and R, but extends beyond them towards the south. This feature is observable in images taken on dates from different seasons over several years, suggesting that it is not a temporary phenomenon. Juxtapositioning this patch with the hypothesized canal L1, we further hypothesize that VP was the command area irrigated by the canal and that the canal itself extended further south. Fourth, a linear feature (L2) whose shape and location suggests that it may have been a wall or bund protecting the settlement area from the river.

By analysing data from fine beam RADARSAT and ENVISAT ASAR, we identified a hitherto unknown buried palaeochannel that skirts to the east of Old Talakadu, cutting across parts of the sand-covered area and parts of the adjoining fields (Rajani et al. 2011) (see Fig. 3.11). The palaeochannel detected on radar data was further enhanced by spectral merging of RADARSAT SAR data of 22 April 2008 with Resourcesat-1 optical LISS-4 data.

Using Cartosat-1 stereo pair imagery, we generated a DEM to understand the terrain undulations of the area of archaeological interest (see Fig. 4.13). We were able to analyse the extent of flooding by simulating different water levels along the bank, and our study revealed two points (see p1 and p2 in Fig. 5.5) which could

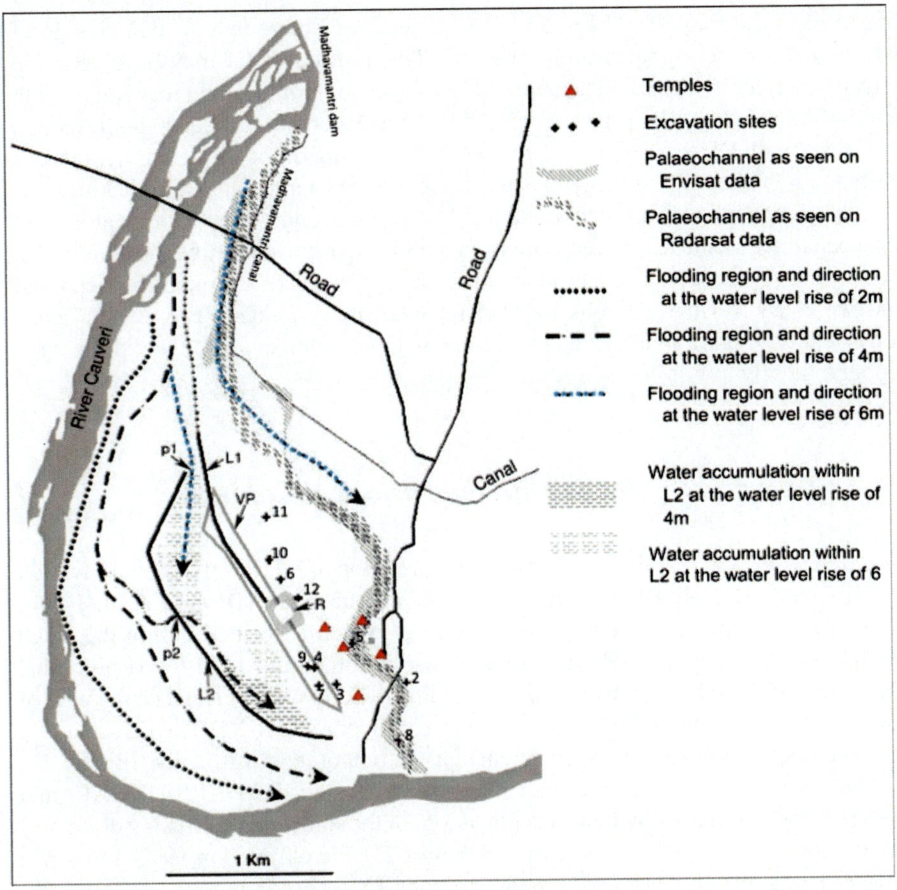

Fig. 5.5 Locations of key findings at Talakadu

indicate breaches (or controlled flooding) into L1 and L2 (Rajangam and Rajani 2017).

5.4.3 Fieldwork

We conducted multiple field visits for this site because its large expanse and sand cover make it time-consuming to survey all points of interest. As always, we preloaded the locations for ground truthing on handheld GPS devices to avoid getting lost among the thick vegetation where there are no identifiable reference points on ground.

The hypothesized canal L1 on ground showed thick vegetation all along, and the topography was noticeably depressed in its northern stretches, suggesting that our interpretation is correct. We found that the northern half of L1 (marked as a dotted line in Fig. 5.5) may have been a river escarpment as there is a sudden change of about 10 feet in topography along the line. This portion of L1 may have been the bank of the river in the past. The feature L2 shows a long ridge along its whole length (kind of morphology illustrated in Fig. 2.4a, b), consistent with our hypothesis that this was a wall. We were unable to find evidence to support or refute our hypotheses for features VP and R, possibly because these are very large and subtle features.

Subsequent field visits were conducted to make ground observations along the palaeochannel identified on radar imagery and flood routes identified on the DEM. Due to heavy rains on the day before this survey, parts of the identified channel that were dry during previous field visits were now flooded (Fig. 3.11). Thus, consistent with our hypothesized palaeochannel, we found evidence that these were topographically low-lying areas.

5.4.4 *Interpreting Findings in the Historical Context*

Two archaeological features support our interpretation of VP. Firstly, past excavation in TK 7 has revealed a bath that was part of an enclosure (Devaraj et al. 1996), giving evidence of water being drawn till TK 7 which is located within the patch of high moisture content VP. Secondly, an inscription dating 1514 CE (Epigraphia Carnatica, 1976) found nearby mentions the grant of cultivated wet land on the west of Kirtinarayana Temple (K), which also comes within or close to VP. The first spatially accurate map of Talakadu was also created as an outcome of this study (Fig. 5.6).

The discovery of the palaeochannel by analysing RADARSAT SAR and ENVISAT ASAR data provided an explanation for the somewhat confusing discovery of river soil 3 m below the surface in trenches of TK 5 (excavated in 1993–94), since the palaeochannel coincides with the location of TK 5 (Fig. 3.11).

The flood layer analysis (Fig. 5.5) has shown the direction of water entry into the Old Talakadu area at different water levels. It also shows an anomaly in the form of sudden turns in water courses (p1 and p2 in Fig. 5.5). These could be breached in either bunds or fortification as they are adjacent to linear features that have been detected in this area.

5.4.5 *Subsequent Archaeological Excavation*

Our experience with Talakadu was unique because in 2009, after our geospatial analysis, a team from Karnataka's Department of Archaeology, Museums and Heritage excavated a trench at a location identified by our analysis. Specifically, excavations were proposed at one location along the hypothesized wall (feature L2, the first priority for excavation) and two locations on the western and eastern sides of the hypothesized reservoir (feature R, the second and third priorities for excavation, respectively), as shown in Fig. 5.6. However, due to limited funds and an abun-

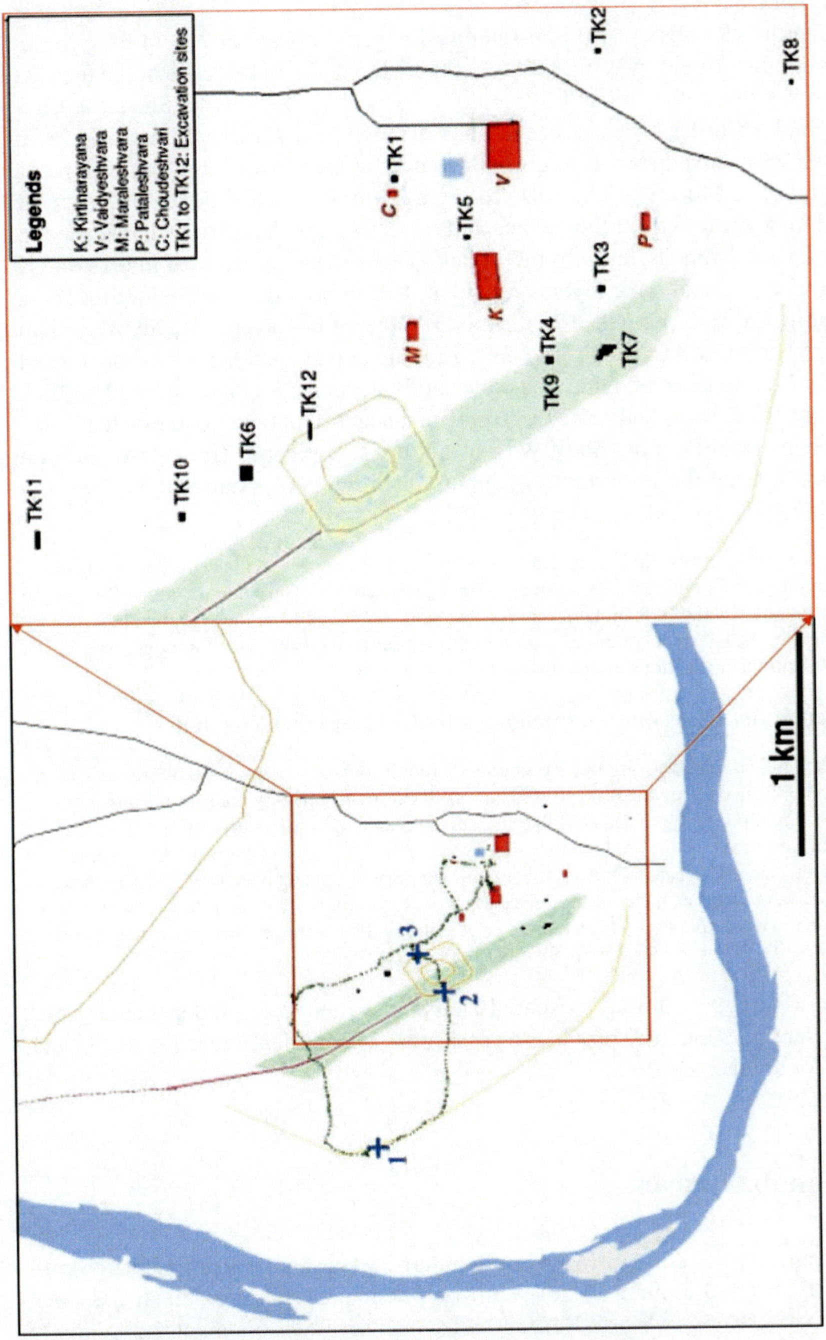

Fig. 5.6 GPS track of the field visit on 7 March 2009 to identify sites for excavation

dance of trees at other locations, excavation was only conducted for the third priority location (named TK 12).

The feature (R) observed in the satellite imagery was about 200 m long in the north–south direction. It was difficult to estimate the width of the sides of the feature, but assuming that it was a bund, it would have been 5–10 m wide. Since the GPS location had a similar error, archaeologists decided to lay a trench that was 23 m long in the west–east direction to give allowance for GPS error. This excavation has been reported in Murthy et al. 2019. To lay a 23 m × 1 m trench, about 3–4 m of sand had to be cleared from the surface and the entire trench had to be barricaded to stop loose sand from sliding into the trench. The trench was divided into two (TK 12 east and west), and three layers were identified in both trenches before reaching natural soil: Layer 1 (humus, 10 cm thick on average), Layer 2 (sandy clay, 1 m thick on average), and Layer 3 (brickbats, 1 m thick on average). TK 12 east showed evidence of being near the edge of a water body as more brickbats were identified. In contrast, TK 12 west had more moisture, consistent with the hypothesis that it lay within the interior of a water body. We now quote a few excerpts from the excavation report highlighting the importance of inputs from the geospatial analysis (Murthy et al. 2019):

> "in layer 2 and 3, river rolled pebbles mixed with small river borne shells were found. These findings indicate that this area was connected to the river source. Thus identification of L-1 as a water canal originating from river Kaveri also gets confirmed from these finds. Therefore identification of L-1 as a water canal and identification of R as the reservoir appears to be correct. Both identifications are confirmed by archaeological sources."

The report discusses our interpretations in detail and then concludes:

> "In this way, the imagery supplied by satellite through remote sensing is highly reliable as they get confirmed by archaeological evidence also. From this imagery many of the unknown facets of the topography of Talakad like the ancient canal system, its origin, its course, the artificial Reservoir, the area of land irrigated by this irrigation system, domestic water supply system, etc. are all revealed. Had not this method of remote sensing been used in the context of Talakad excavations, many of the subtle aspects of the cultural legacy of this town would have remained unexplained. This technology, for sure, will help the archaeologists to unravel many more facets of our legacy not visible to the naked eye."

Having findings from our geospatial analysis ratified by an independent group using conventional methods provides a great sense of confidence in the value of such analyses for archaeology.

5.5 Mahabalipuram

Mahabalipuram was a port city of the South Indian Pallava Dynasty during the seventh century CE, located around 60 km south of Chennai in Tamil Nadu. It has been classified as a UNESCO World Heritage Site, with several monuments built largely between the seventh and ninth centuries CE. These monuments include cave temples, monolithic free-standing shrines known as *rathas* (chariots), sculpted reliefs, and structural temples (Harle 1986). Early European explorers gave Mahabalipuram the nickname Seven Pagodas, because the summits of seven temples could be observed from the sea—they served as a landmark for navigation. With changes to sea levels and coastlines, the view from the sea is different today, and only one temple (the Shore Temple) is visible from the sea. There are several speculations: one hypothesis is that the Shore Temple was one of the Seven Pagodas and that the remaining six are now under water as sea levels rose. This speculation is supported by the fact that the ruins of several temples have been found underwater near the shore (Sundaresh et al. 2004). Another hypothesis is that if the Shore Temple (which has two towers) is counted as two pagodas, then the cluster of five monuments called Pancha Rathas (see X in Fig. 2.28), which are the next closest set of temples to the shore, would add up to Seven Pagodas. We studied this site to examine these competing hypotheses or to propose an alternate hypothesis (Rajani and Kasturirangan 2013).

5.5.1 Historical Spatial Records

Our investigation was triggered by the chance finding of a Dutch portolan chart (maritime map) dated 1670, in the archives of the Royal Geographical Society. Although the chart is entitled "Coast of Malabar", it unmistakably refers to the stretch of India's south-east coast from Pulicat to Rameswaram, as well as some parts of northern Sri Lanka. (In classical European accounts, the word Malabar was often used to refer to South India.) This chart depicts relatively permanent features near the coast such as forts, temples, and hills that were visible from sea at the time, and place names can be identified with their modern equivalents (see Fig. 4.7). Mahabalipuram is marked as Seven Pagodas, and seven distinctively shaped temples are distributed along a shoreline whose peninsula-like shape is very different to the shape of the modern shoreline (see Fig. 2.28b).

5.5.2 Preliminary Geospatial Analysis

Each of the Seven Pagodas marked on the portolan chart has a specific shape and position along the shoreline. For instance, the Shore Temple can be easily recognized by its twin towers and is depicted as standing in the water. Today, this temple stands well inland, which suggests that the shoreline must have been further inland. (Note that this immediately casts doubt on the first hypothesis noted earlier, which requires the modern sea level to be higher.) Next, we compared ground photographs of temples with the portolan chart to identify similarly shaped temples, keeping in mind their actual location and their relative location as depicted on the portolan chart. For instance, the temple in the portolan chart marked with a flat roof and located in the middle of the peninsula corresponds well with the Olakkanatha Temple (Fig. 2.28), which sits on top of a rock near a lighthouse and is often called the lighthouse temple.

We identified each of the remaining temples depicted in the map in a similar way. These associations also cast doubt on the second hypothesis noted earlier, since they imply that the Pancha Rathas would have been submerged because they are not marked on the map. Since the Pancha Rathas do not show signs of erosion by water, they may in fact have additionally been covered by sand.

5.5.3 *Geospatial Analysis*

Since we believed that the shoreline was further inland, we used imagery from the Corona satellite as well as multispectral imagery from LISS-3 and LISS-4 to identify strand lines that show trends in the movement of shorelines. Google Earth imagery across multiple dates also provided inputs on the shoreline changes and helped in matching the seven structures with their depictions on the portolan chart.

In addition, we conducted topographical analysis with a DEM to see if we could recreate the shape of the shoreline shown in the chart by simulating sea level rise. Note that such a simulated shoreline cannot be expected to have the same shape as the shoreline depicted in the portolan chart for two reasons. First, given the mapping methods available at the time, the chart is likely to have significant spatial inaccuracies. Second, the map is over three centuries old, and coastal processes are likely to have impacted the coastline and topography. Nevertheless, we created 3D skeletal models of our hypothesized Seven Pagodas and placed them (with their accurate heights) on a virtual 3D terrain. We then created a water layer and simulated its rise metre by metre until the hypothesized temples were close to the shoreline (Fig. 5.7).

As expected, the generated shoreline does not correspond exactly to the shape of the portolan chart's shoreline, but there are important similarities. For example, the peninsula shown in the portolan chart corresponds to an elevated area. Figure 5.7 shows the simulated view of the site and Seven Pagodas as seen from the (elevated) sea. Figure 5.7a shows the view from the south, Fig. 5.7b shows from the east, and Fig. 5.7c shows from the north. While all Seven Pagodas are seen in the latter view, some of these structures are obscured by topography in the other views.

5.5.4 *Fieldwork and Interpreting Findings in the Historical Context*

In this study, fieldwork was only needed to take ground photographs and GPS readings of each of the temples. In addition, one could identify additional shoreline features and collect shoreline material (shells/aquatic fauna along strandlines) for dating, but we have not explored this so far.

Our study has provided a novel explanation for Mahabalipuram's mysterious name Seven Pagodas by identifying seven temples that correspond to seven structures

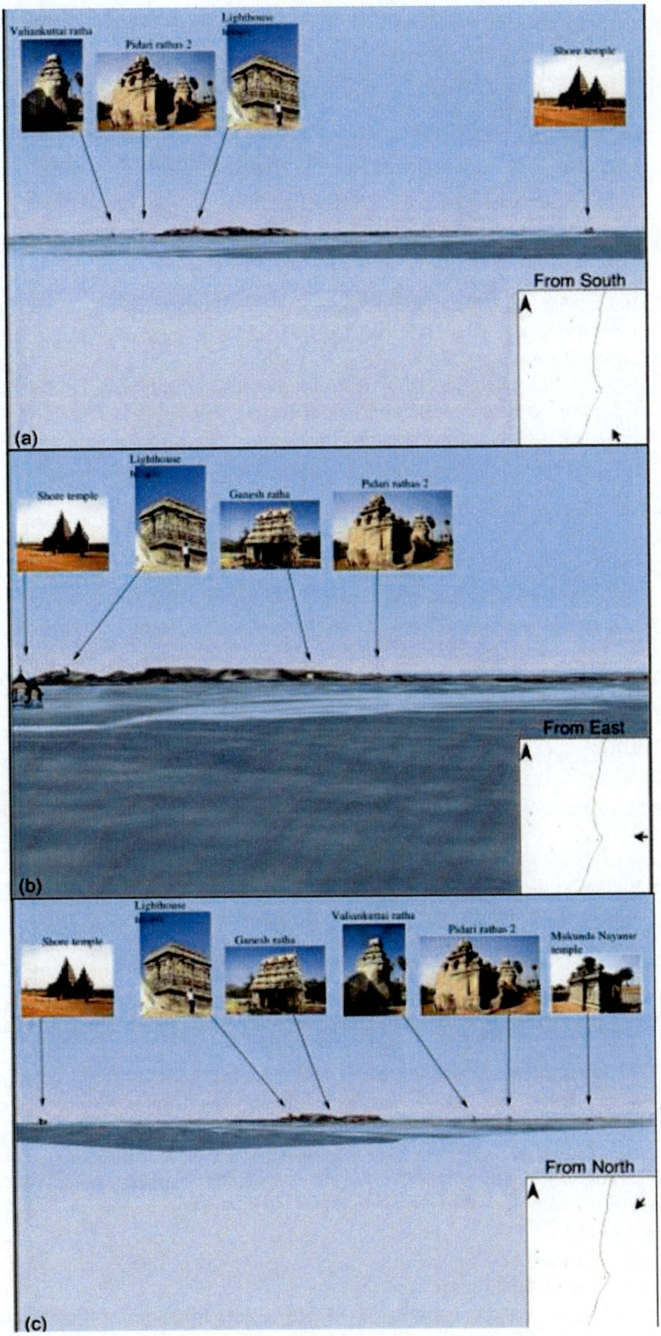

Fig. 5.7 Simulated views of Mahabalipuram from the sea with raised sea level **a** view from the south, **b** from the east, **c** from the south

marked on the portolan chart from 1670. This work makes two important contributions to the study of other coastal sites. First, it has demonstrated the potential value of spatial information recorded in portolan charts. Second, it has demonstrated the effectiveness of virtual GIS tools to study spatial contexts of coastal sites.

References

Asher F (2018) India, Magadha, Nalanda: ecology and a premodern world system. In: Sharma A (ed) Records, recoveries, remnants and inter-Asian interconnections: decoding cultural heritage. ISEAS–Yusof Ishak Institute, pp 51–69

Beal S (trans) (1884) Si-yu-ki Buddhist Records of the Western World, Vols. I & II London, Kegan Paul. http://archive.org/details/siyukibuddhistr00bealgoog. Accessed 17 Apr 2020

Beal S (trans) (1914) The life of Hiuen-Tsiang by the Shaman Hwui-Li, London. http://archive.org/details/lifeofhiuentsian030569mbp. Accessed 17 Apr 2020

Buchanan F (1807) A journey from Madras through the countries of Mysore, Canara, and Malabar…vol 1, Chap II, London. https://archive.org/details/journeyfromMadr1Hami/page/iii/mode/2up. Accessed 17 Apr 2020

Buchanan F (1812) BL: MSS EURD87, p. 127

Carlleyle ACL (1874) Annual Report for the Year 1871–72 (Delhi by J.D. Beglar) and (Agra by A. C. L. Carlleyle) under Major General A. Cunningham, vol IV (Calcutta: Archaeological Survey of India), p 23

Cunningham A (1871) Archaeological Survey of India: four Reports made during the years 1862-63-64-65. 1(Plate XVI) Simla (https://archive.org/details/report00cunngoog)

Das S, Rajani MB (2020) A geospatial study of archaeological remains at Halebidu: An integrative approach to identify unexplored features. Journal of Indian Society of Remote Sensing, submitted, unpublished

Devaraj DV, Murthy AVN, Swamy MSK (1996) Excavations at Talakad. Directorate of Archaeology and Museums, Mysore

Epigraphia Carnatica (1976) Vol V, (Revised), Mysore

Gole S (1989) Indian Maps from earlier times to the advent of European surveys. Manohar Publications, New Delhi, p 200

Gupta E, Das S, Rajani MB (2017) Archaeological exploration in Srirangapatna and its environ through remote sensing analysis. J Indian Soc Remote Sens 45:1057–1063. https://doi.org/10.1007/s12524-017-0659-9

Harle JC (1986) Art and architecture of the Indian subcontinent, the pelican history of art. Penguin, England

Hodges W (1793) Travels in India during the Years 1780, 1781, 1782 and 1783. J. Edwards, London

Hunter J (1792) Garden Gate, Laul Baugh, Seringapatam. http://www.bl.uk/onlinegallery/onlineex/apac/other/019xzz000007683u00006000.html. Accessed 17 Apr 2020

Iyer M, Nagendra H, Rajani MB (2012) Using satellite imagery and historical maps to investigate the original contours of Lalbagh Botanical Garden. Current Science 102(3) 10 Feb:507–509

Jackson VH (ed) (1922) Journal of Francis Buchanan (Patna and Gaya Districts). JBihar Orissa Res Soc VIII(III & IV):263

John JK (1997) The mapping of Hindustan: a forgotten geographer of India, Joseph Tieffenthaler (1710–1785). Proc Indian Hist Congr 58:400–410

Khan MA (1993) Glimpses of the administration of Agra under Muhammad Shah (Based on news reports from the city). Proc Indian Hist Congr 54:200–207

Koch E (1986) The Zahara Bagh (Bagh - i - Jahanara) at Agra. Environ Des J Islamic Environ Des Res Centre 2:30–37

Koch E (1997) Mughal Palace Gardens from Babut to Shahjahan (1526-1648)'. Muqarnas 14:143–165

Koch E (2005) The Taj Mahal: Architecture, Symbolism, and Urban Significance. Muqarnas 22:128–149

Koch E (2006) The Complete Taj Mahal, first, paperback edn. Thames and Hudson Ltd, London, p 22

Koch E (2008) Mughal Agra: A Riverfront Garden City. In: Jayyusi SK et.al (eds) The City in the Islamic World, vol 1. Brill, Leiden and Boston, pp 555–588

Koch E (2019) Palaces, gardens and property rights under Shah Jahan: architecture as a window into Mughal legal custom and practice. In: Koch E, Anooshahr A (eds) The Mughal empire from Jahangir to Shah Jahan: art, architecture, politics, law and literature. Marg Foundation, Mumbai, pp 196–219

Koch E, Losty JP (2017) The riverside mansions and tombs of Agra: new evidence from a panoramic scroll recently acquired by the British library. Electron Br Libr J 9:1–33

Laidley JW (trans) (1848) The pilgrimage of Fa Hian, Calcutta. English trans. from French edition of Rémusat A, Klaproth JV, Landresse C (1836) Foe Koue Ki, Paris. https://archive.org/details/pilgrimagefahia00rmgoog. Accessed 16 Apr 2020

Latif KBSM (1896) Agra historical & descriptive, with an account of Akbar and his court. Calcutta Central Press, Calcutta

Misra BN (1998) Nalanda, 3 vols. B. R. Publishing, Delhi

Moienuddin M (2000) Sunset at Srirangapatam after the death of Tipu Sultan. Orient Longman Limited, Hyderabad, p 39

Moreland WH, Geyl P (1925) Jahangir's India (The Remonstrantie of Francisco Pelsaert) Translated from Dutch. W. Heffer & Sons Ltd., Cambridge

Murthy MSK, Gopal R, Gangadhara TS (2019) Archaeological excavations at Talakad 2006-2010 (2). Department of Archaeology, Museums and Heritage, Karnataka, India

Peck L (2011) Agra: the architectural heritage. Roli Books Private Limited, New Delhi

Phuoc LH (2010) Buddhist Architecture ([S.I.]: Grafikol)

Rajangam K, Rajani MB (2017) Applications of geospatial technology in management of cultural heritage sites—potentials and challenges for the Indian region. Navalgund RR, Korisettar R (eds) Geospatial techniques in archaeology. Curr Sci Special Section 113(10) 25 Nov:1948–1960

Rajani MB (2016) The expanse of archaeological remains at Nalanda: a study using Remote Sensing and GIS'. Archives of Asian Art, Duke University Press 66(1)Spring:1–23

Rajani MB, Bhattacharya S, and Rajawat AS (2011) Synergistic application of optical and radar data for archaeological exploration in the Talakadu Region, Karnataka. J Indian Soc Remote Sens Springer 39(4)Dec:519–527. https://doi.org/10.1007/s12524-011-0102-6

Rajani MB, Das S (2018) Archaeological remains at Nalanda: a spatial comparison of nineteenth century observations and the protected world heritage site. In: Sharma A (ed) Records, recoveries, remnants and inter-Asian interconnections: decoding cultural heritage. ISEAS–Yusof Ishak Institute, pp 239–256

Rajani MB, Kasturirangan K (2013) Sea level changes and its impact on coastal archaeological monuments: seven pagodas of Mahabalipuram, a case study. J Indian Soc Remote Sens 41:461–468. https://doi.org/10.1007/s12524-012-0210-y

Rajani MB, Patra SK (2009) Application of satellite image processing techniques for Talakadu, a unique archaeological landscape in India', Photo-Interpretation, ESKA, Paris, France 45(4):168–175

Rajani MB, Rajawat AS, Murthy MSK, Kamini J, Rao S (2012) Demonstration of the synergy between multi-sensor satellite data, GIS and ground truth to explore the archaeological site in Talakadu region in South India. J Geomatics Indian Soc Geomatics 6(1)Apr:37–41

Rice BL (1897) Plan of Seringapatam. Mysore: a gazetteer compiled for Government 2:319. http://britishlibrary.georeferencer.com/maps/5ba07906-a2eb-5ba3-9760-d981870dd528/. Accessed 17 Apr 2020

Sivaramkrishna S (2005) The curse of Talakad. Rupa & Co., New Delhi

Smith VA (1916) Travels in Mughal Empire (1656–1668 A.D.) by Francois Bernier. Translated on the basis of Irving Brock's version and annotated by Archibald Constable (1891), 2nd edn. Oxford University Press, London, Edinburgh, Glasgow, New York, Toronoto, Melbourne, Bombay: Humphrey Milford

Stewart ML (1989) Nalanda Mahavihara: a study of an Indian Pala period buddhist site and british historical archaeology, 1861–1938, BAR International Series 529. B.A.R, Oxford

Suganya K, Rajani MB (2020) Riverfront gardens and city walls of Mughal Agra: a study of their locations, extent and subsequent transformations using remote sensing and GIS. South Asian Studies. https://doi.org/10.1080/02666030.2020.1721119

Sundaresh Gaur AS, Tripati A, Vora KH (2004) Underwater investigations off Mahabalipuram, Tamil Nadu, India. Curr Sci 86(9):1231–1237

Swamy LN (2010) Srirangapattana Fort through the ages origin, development and the fall of the Fort. In: Gopal R (ed) Tipu Sultan the tiger of Mysore. Directorate of Archaeology and Museums, Mysore

Takakusu J (1876) A record of the buddhist religion as practiced in India and the Malay Archipelago (A.D. 671–675) by I-tsing (Nan-hae-ki-kwei-niu-fa-ch'uen). Clarendon Press, Oxford. https://archive.org/details/arecordbuddhist00takagoog. Accessed 17 Apr 2020

Twining WHJ (1893) Travels in India a hundred years ago with a visit to the United States being notes and reminiscences by Thomas Twining. James R Osgood, Mcllvaine & Co., London

Valdiya KS (2008) Sinking of ancient Talakad temples on the Kaveri Bank, Mysore Plateau, Karnataka. Curr Sci 95(12):1675–1676

Chapter 6
Site Protection Boundaries: A Double-Edged Sword

Preamble

The protection and preservation of built heritage are particularly challenging in densely populated developing countries such as India, where the rapid expansion of towns, industries, and transportation networks places an immense premium on land. Local, national, and international authorities tasked with protecting cultural heritage sites have sought to limit development in the vicinity of these sites. Simple definitions of boundaries that demarcate the vicinity are important for two reasons. First, after identifying an individual structure (or a set of structures) as built heritage, protection boundaries can be defined straightforwardly. Second, simple definitions can streamline the resolution of actual or potential conflicts and impose clear mandates on authorities to protect structures within these boundaries. On the other hand, *overly* simplistic definitions may exclude known or hitherto undiscovered remnants of built heritage (perhaps camouflaged by modern landcover) from the protected region. Note that the second advantage of simple definitions is less compelling given the widespread adoption of GIS tools that can define incredibly complex boundaries (e.g. gerrymandered political boundaries, Shashidhar 2019) and GNSS tools that can accurately indicate whether a location is within or outside these boundaries. Hence, a simple definition of boundaries trades the advantage of rapidly defining protected regions against the risk of exposing excluded structures to unfettered development.

In the first part of this chapter (Sect. 6.1), we will explore this trade-off in the context of the definition of boundaries given by the Archaeological Survey of India (ASI) and UNESCO's World Heritage Convention (WHC). In the rest of this chapter, we will propose two policy recommendations that would greatly assist in preserving the fragile remnants of our built heritage: using geospatial analysis to create more nuanced and meaningful site protection boundaries (Sect. 6.2) and creating a national-level geospatial database for built heritage (Sect. 6.3).

M. B. Rajani, *Patterns in Past Settlements: Geospatial Analysis of Imprints of Cultural Heritage on Landscapes*, Springer Remote Sensing/Photogrammetry,
https://doi.org/10.1007/978-981-15-7466-5_6

6.1 ASI and WHC Boundaries

The ASI was created in 1861 and tasked with undertaking "a complete search over the whole country, and a systematic record and description of all architectural and other remains that are either remarkable for their antiquity, or their beauty or their historical interest"[1]. Today, the ASI looks after over 3600 sites, and the Department of Archaeology of each state typically takes care of a few hundred sites.

The sites protected by ASI and the states are governed by the Ancient Monuments and Archaeological Sites and Remains (Amendment and Validation) Act, 2010, popularly known as the AMASR Act.[2] This act defines protected area as any archaeological site and remains declared to be of national importance by or under the act. Additional land can be included for fencing, covering, or otherwise preserving the remains and providing adequate access to them. These judgements are typically made based on surveys of remains that are visible from the ground. Once the protected area is determined, the act provides simple definitions of prohibited and regulated areas, which, respectively, extend to 100 m and an additional 200 m in all directions from the protected area. Thus, the prohibited and regulated regions form two concentric annuli around the protected area, and ISRO has used automated GIS tools to digitize these regions for ASI on the Bhuvan geoportal.[3] Further, an associated online decision support system named SMARAC[4] has been created to efficiently process requests for clearances to develop plots of land as per this classification.[5] While this is an undeniably valuable service, the simple definition of boundaries has excluded unprotected built heritage structures in the vicinity of protected structures and elevated the risk they face (see Sect. 6.3).

UNESCO's World Heritage Convention adopts a different definition of boundaries. The operational guidelines state that the delineation of boundaries is "an essential requirement in the establishment of effective protection" of heritage properties and necessitates marking so-called core and buffer zones.[6] The present guidelines (Chapter II.F, para 99–107) for drawing cultural property boundaries require that boundaries are drawn to include "all those areas and attributes which are a direct tangible expression of the Outstanding Universal Value of the property". Further, the WHC recognizes the elevated threat posed by changing social and economic conditions to landscapes that possess objects of heritage value. Thus, boundaries are

---

[1]https://asi.nic.in/about-us/history/. Accessed on 25 May 2020.

[2]http://www.nma.gov.in/documents/10184/0/AMASR_Act2010_Gazette_Notification.pdf/7bbf8c39-f561-49ab-b844-c3f7051238c3. Accessed on 24 April 2020.

[3]Heritage sites and monuments. https://bhuvan-app1.nrsc.gov.in/culture_monuments/. Accessed on 24 April 2020.

[4]SMARAC Decision Support System (DSS). https://www.nrsc.gov.in/readmore_dss?language_content_entity=en. Accessed on 24 April 2020.

[5]NOC Online Application and Processing System, National Monuments Authority, Ministry of Culture. http://nmanoc.gov.in/. Accessed on 24 April 2020.

[6]The Operational Guidelines for the Implementation of the World Heritage Convention. Chapter II.F, para 99–107. https://whc.unesco.org/document/178167. Accessed on 24 April 2020.

allowed to include "those areas which in the light of future research possibilities offer potential to contribute to and enhance such understanding". When authorities wish to propose a site for inclusion on the list of World Heritage Sites (which contains 869 cultural properties to date, out of which 30 are in India), this definition offers an opportunity to include regions where further research is necessary to determine the extent of surviving cultural remains.

It is concerning that UNESCO's website lists no maps for five of the WHC sites from India. Over the years, the WHC has started to insist that all "nomination dossiers" must include maps indicating monument zones, protection boundaries, core and buffer zones, and habitation sites contextualized with topographical and cadastral maps. Without such information, sites are not accepted for the WHC's nomination. Today, the dossiers submitted for Indian sites include such maps, created by GIS software. As an example, compare the map of Hampi when it was first proposed as a World Heritage Site in the 1980s (Fig. 6.1) with the map from 2012 created by GIS (Fig. 6.2) (Rajangam and Rajani 2017). Nevertheless, these maps are typically prepared by integrating spatial data from just three main sources: topography (which in case of India come from SOI), cadastral (local authority for revenue and land records), and heritage site plans. In most cases, the latter only consists of architectural plans of protected structures. Crucially, the dossiers do not include cultural heritage features in the larger landscape that are detectable by geospatial analysis as described in this book. As the following case studies show, such features often fall outside the protected boundaries.

6.2 Importance of Using Geospatial Analysis While Creating Site Protection Boundaries

While India has successfully listed several properties as World Heritage Sites[7] and has nominated further sites,[8] it has thus far failed, in most cases, to propose core zone boundaries beyond the area currently protected by ASI. The WHC recognizes that some countries lack economic, scientific, and technological resources to develop an effective and permanent system of protection in accordance with modern scientific methods.[9]

These challenges should not apply to India, and we argue that geospatial analysis is essential in defining appropriate site protection boundaries. To support this argument, we will consider three sites that vary in the type and duration of protection accorded to them: Sarnath (part of the World Heritage Sites Tentative List since 1998) and two sites inscribed as World Heritage Sites (Bodhgaya in 2002 and Nalanda in 2016). All three sites are similar in size, location (away from big cities), and with

[7] World Heritage List. http://whc.unesco.org/en/list/. Accessed on 24 April 2020.

[8] Tentative Lists. http://whc.unesco.org/en/tentativelists/state=in. Accessed on 24 April 2020.

[9] Convention Concerning the Protection of the World Cultural and Natural Heritage. http://whc.unesco.org/en/conventiontext/. Accessed on 24 April 2020.

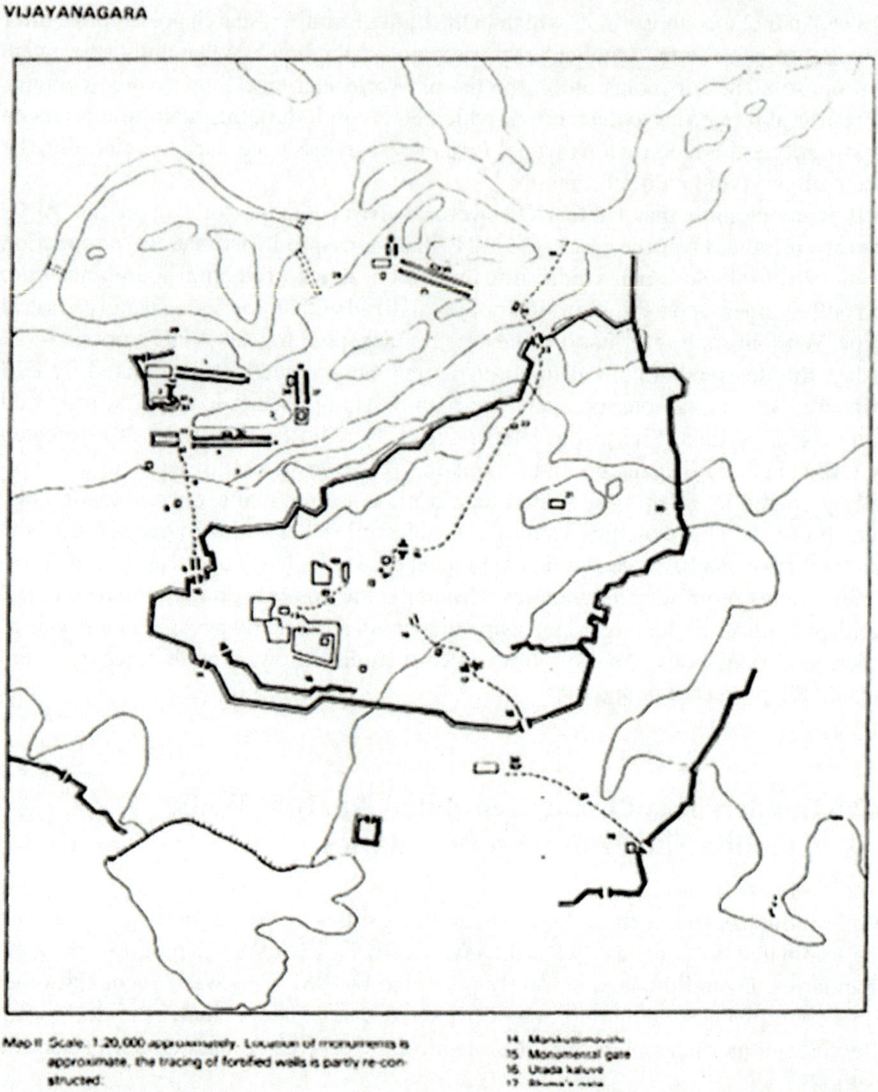

Fig. 6.1 World Heritage boundary proposed in 1982 for the site of Hampi (reproduced from Rajangam and Rajani 2017)

a high tourist footfall (all of them are important Buddhist sites). Unsurprisingly, development activities are highly pronounced near popular sites and are spurred by the attainment of UNESCO's World Heritage tag. At each site, we consider the (in)sufficiency of boundaries according to the ASI and UNESCO definitions by examining the threat to nearby cultural remains from development activities. We demonstrate that such remains could have been identified using geospatial analysis

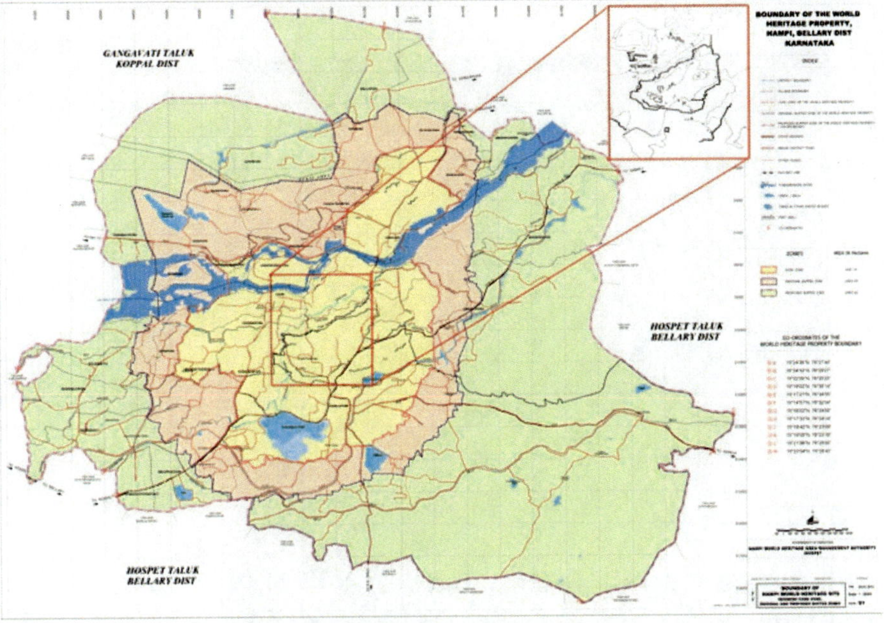

Fig. 6.2 The core (yellow) and buffer (green) zones of Hampi World Heritage Site with an overlay of the 1982 proposed core zone (red box). Background map https://whc.unesco.org/document/119806. Accessed on 31 May 2020

and included within the core or buffer zone for their "future research possibilities" as per the WHC's guidelines.

6.2.1 Sarnath

Sarnath, near Varanasi (Uttar Pradesh), has been on UNESCO's Tentative List since 1998. The Buddha is believed to have first preached in Sarnath, making it one of the four holiest Buddhist sites (alongside Lumbini in Nepal where he was born, Bodhgaya in Bihar where he attained enlightenment, and Kushinagar in Uttar Pradesh where he attained Nirvana). Sarnath is located near the confluence of the rivers Ganga and Varuna. The protected area has two distinct plots: an approximately 60,000 m^2 complex of excavated structures (which includes Dhamekh and Jagat Singh stupas) and the approximately 13,500 m^2 Chaukhandi stupa located 750 m further south (see Fig. 6.3). A geospatial analysis of the region surrounding the protected area using satellite images has identified cropmarks indicating an ancient settlement covering an area of 3.7 km^2 (Gupta et al. 2017; Asher 2020). The layout is rectangular, flanked by water bodies that are now dry, with sides oriented to the cardinal directions (Fig. 6.3).

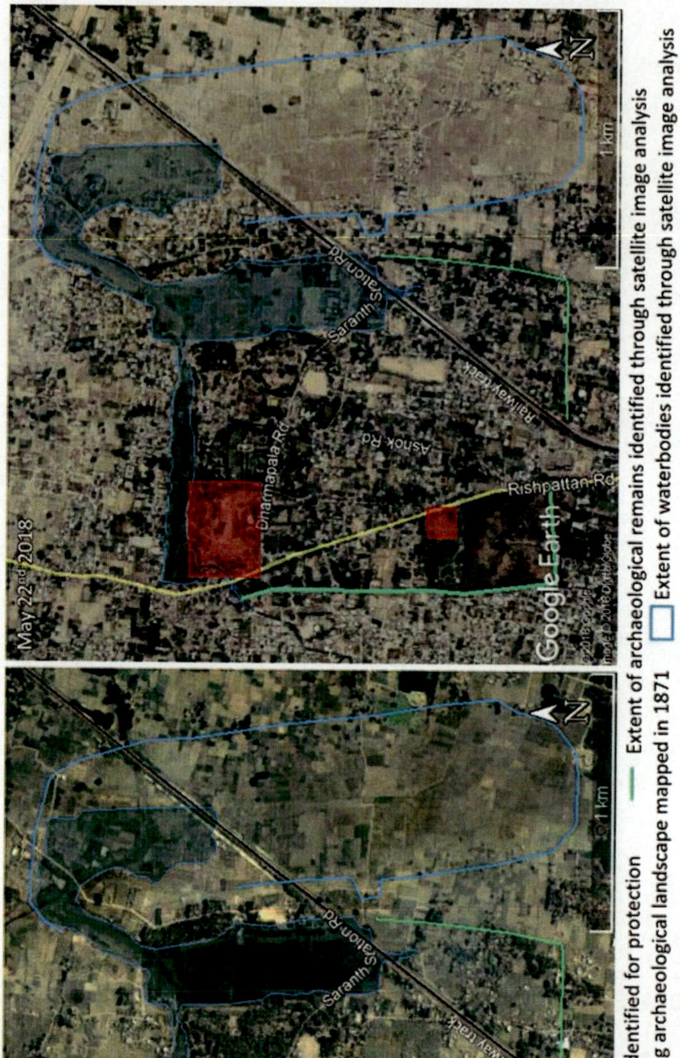

Fig. 6.3 Area identified for protection at Sarnath in its spatial context. Note the increased number of concrete structures from 2002 (**a**) to 2018 (**b**)

Alexander Cunningham visited the site in 1835–36 and reported on the initial explorations (Cunningham 1871). Subsequent excavations, including some as recently as 2014,[10] have uncovered monasteries, stupas (Dhamekh, Dharamrajika and Chaukhandi), and antiquities dating from the third century BCE to the twelfth century CE.[11] Cunningham's report of 1871 includes a map showing the larger archaeological landscape that contained three large interconnected tanks and many mounds of anthropogenic shape (Fig. 4.6).[12] By georeferencing this map, we find that the 1.4 km^2 area Cunningham surveyed in 1871 lies within the northern half of the 3.7 km^2 rectangular settlement. This archaeological landscape has been fragmented by a major road (which has developed over time from a small path in Cunningham's sketch) and a railway line from Varanasi to Gorakhpur. Further exploration within this rectangular area using non-destructive geophysical techniques has high potential for identifying buried features of architectural interest, some of which could be worth excavating and conserving. However, Fig. 6.3 makes it clear that several buildings have cropped up between 2002 and 2018 at multiple locations within this rectangle and near the water bodies (but outside the area protected by ASI), potentially destroying such remains. It is therefore imperative that Sarnath's larger expanse receives protection status. Ideally, the entire 3.7 km^2 rectangular region and the water bodies should be included for protection when the proposal is put forward.

6.2.2 *Bodhgaya*

Bodhgaya's main shrine, the Mahabodhi Temple, was inscribed as a World Heritage Site in 2002.[13] It marks the location where Buddha is said to have attained enlightenment. It is therefore among the four most important sites connected to Buddha's life, and it has been a major pilgrimage destination for Buddhists for well over two thousand years. The main shrine was constructed in the fifth/sixth century CE, although its present form dates to the end of the nineteenth century. The larger settlement had a square monastery to the north that was visible in the nineteenth century. Several other remains had formed into mounds and were surveyed and mapped by Cunningham in 1892. This map has been georeferenced, and all its features have been identified on satellite imagery (Fig. 6.4), although the clarity of their contours has diminished from 2003 to 2018 as several modern buildings have been constructed. This includes

[10]Sarnath Circle, Archaeological Survey of India, https://www.asisarnathSarnathcircle.org/excavation-exploration.php. Accessed on 21 April 2020.

[11]http://asi.nic.in/sarnath/. Accessed on 21 April 2020.

[12]Cunningham (1871), Archaeological Survey of India: four reports made during the years 1862-63-64-65. 1 (Plate XXXI) Simla (https://archive.org/details/fourreportsmade00cunngoog/page/n221). Accessed on 21 April 2020.

[13]Mahabodhi Temple Complex at Bodh Gaya. http://whc.unesco.org/en/list/1056/multiple=1&unique_number=1231. Accessed on 24 April 2020.

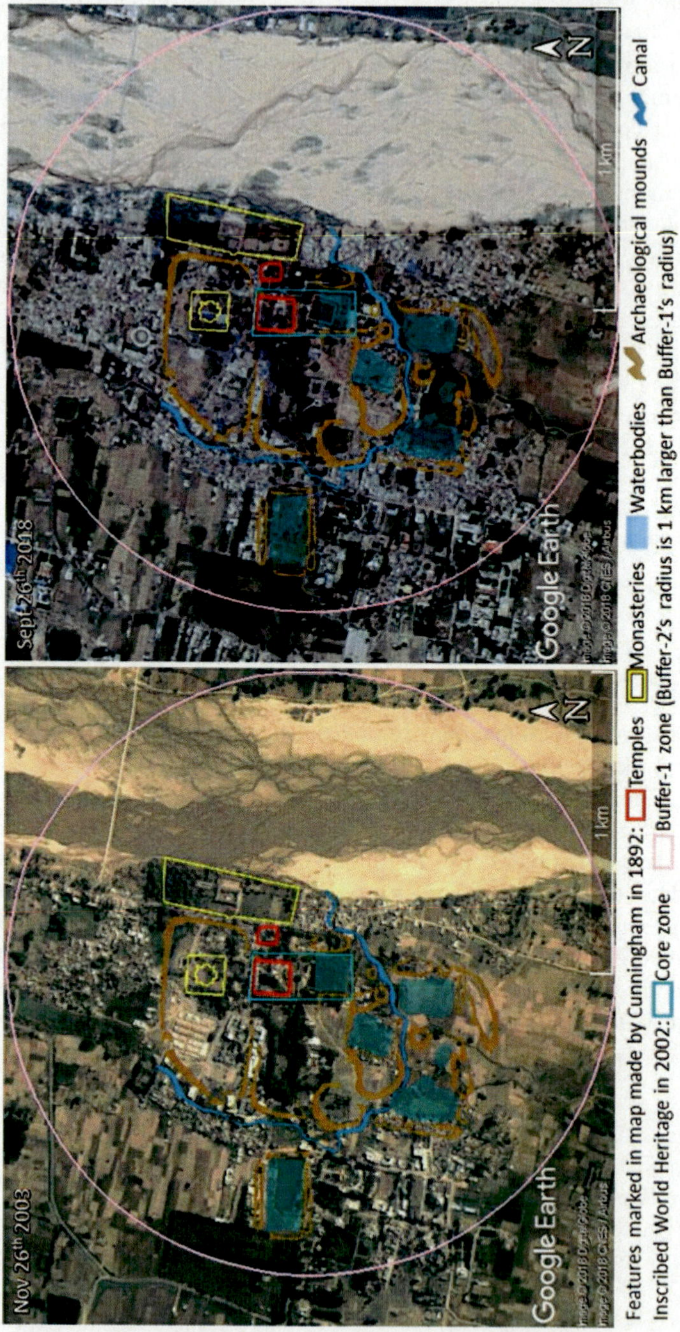

Fig. 6.4 Property of Bodhgaya inscribed as a World Heritage Site in 2002 in its spatial context. Note the increased number of concrete structures from 2003 (**a**) to 2018 (**b**)

construction even within the region marked Buffer-1 in the map for Bodhgaya on UNESCO's website.[14]

When we conducted a field survey in 2018, accessibility and mobility in the larger landscape were very restricted, especially during the run-up to festivals and during vacations. When the site was inscribed as a World Heritage in 2002, only the temple complex and the tank to its south were included in the core zone. Ideally, the core zone should have been expanded to include all the mounds, tanks, canal, and monasteries marked in Fig. 6.4, to protect them from the development that has subsequently taken place. It may still be worth extending the protected boundary to include the larger landscape in the interest of avoiding further damage to these remains. Although the temple is not under the protection of the ASI, it might be possible for the ASI to assume authority over the larger landscape.

6.2.3 Nalanda

Nalanda was inscribed as a World Heritage Site in 2016,[15] and Fig. 6.5 shows the core and buffer zones. The inscribed core zone (0.23 km^2) corresponds to the area owned by ASI, which includes all the protected structures (temples and monasteries). The boundary of the buffer zone (0.58 km^2) is defined by cadastral lines about 300 m from core zone boundary, skirting along where ASI's regulated area ends (although this buffer is pinched to the north and south near the villages of Baragaon and Muzaffarpur, respectively) (Fig. 6.5). Our research (Rajani 2016; Rajani and Das 2018; summarized in Sect. 5.1) suggests that the site spanned approximately 7.25 km^2. (Normalizing this to align with land parcel boundaries would yield an area of 9.8 km^2 to protect as shown in Fig. 6.6). Even prior to our research (see "unprotected archaeological mounds" marked in Fig. 6.6), it has long been known that several mounds in the vicinity have archaeological potential (Cunningham 1871; Asher 2015; Misra 1998; Stewart 1989), but these have been excluded from the core and buffer zones. Interest in Nalanda is certain to increase with World Heritage recognition, and there is a significant risk that these expressions of cultural heritage in the spatial context of Nalanda will suffer the same fate as the structures near Bodhgaya unless they are included within the protected regions.

---

[14]Mahabodhi Temple Guide Map (2002) http://whc.unesco.org/document/103331. Accessed on 21 April 2020.

[15]Archaeological Site of Nalanda Mahavihara at Nalanda, Bihar. https://whc.unesco.org/en/list/1502. Accessed on 21 April 2020.

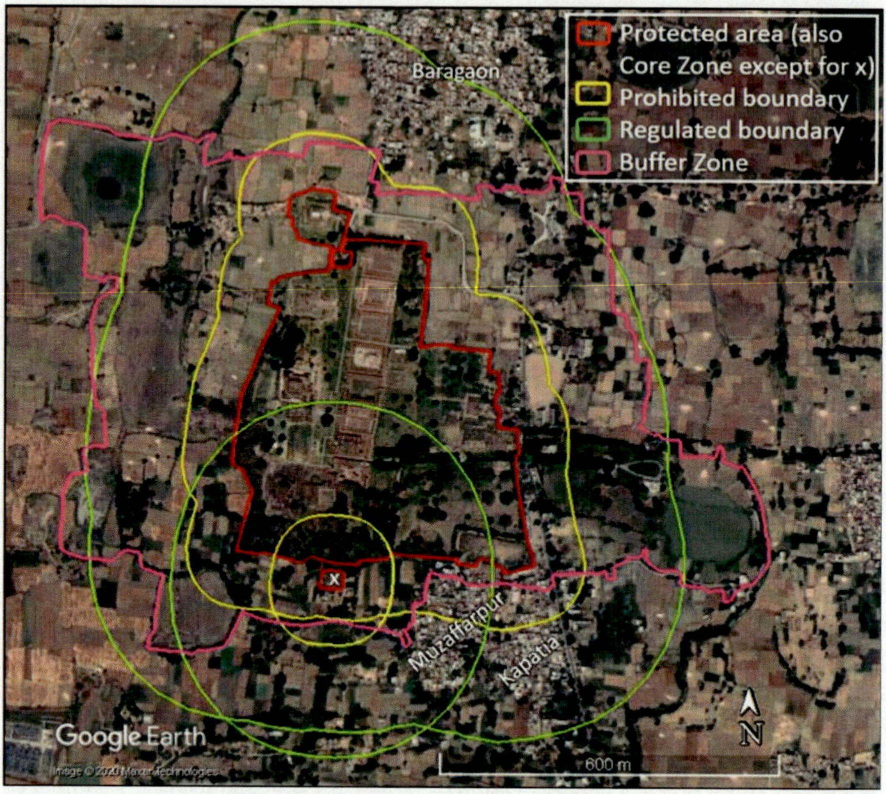

Fig. 6.5 Protection boundaries around the site of Nalanda

6.3 National-Level Geospatial Database for Built Heritage

During the nineteenth century, several public work projects caused significant damage to archaeological landscapes. Examples include the fragmentation of the spatial contexts by roads and railway lines of Sarnath (Fig. 6.3) and Agra (Fig. 5.3). Since the pace of such activities is expected to accelerate, it is critical to have a geospatial database of built heritage available for planners. Unlike a database such as Bhuvan, this resource must include not just confirmed sites, but also potential locations of built heritage identified through geospatial analysis. When new infrastructure projects are conceived, planners must mandatorily consult this database to ensure that they respect the protection boundaries around confirmed and potential built heritage during the planning phase. In the absence of such a database, the existence of built heritage can be overlooked.

As an example, consider Chikkajala, a village about 15 km north of Bangalore, has prehistoric sites protected by the ASI to the south of the village. To its north is an unprotected fort (estimated to be at least 200 years old) with a temple, a tank, a

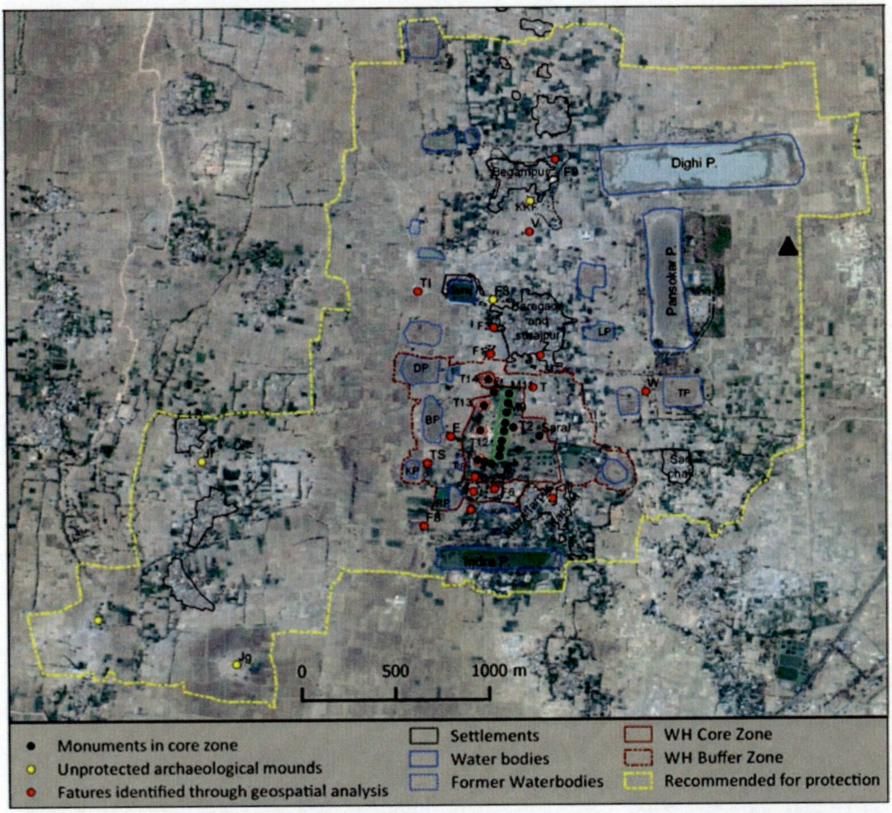

Fig. 6.6 Spatial context of Nalanda and the region recommended for protection (dotted yellow boundary)

pillared pavilion, and a residential space within the 2-acre enclosure. A chronological sequence of four satellite images in Fig. 6.7 shows how the fort has gradually succumbed to development (Gupta et al. 2017). In 2004, the highway (NH-7) was a single carriageway, with the fort located at a safe distance. In 2010, the highway was expanded into a dual carriageway, with a service road abutting the ill-fated fort. Thereafter, the National Highway Authority of India was granted permission to break a segment of the fort's outer wall to widen the highway and accommodate a flyover. The sequence of images clearly shows how rapidly the site became overgrown (disintegrating the temple, tank, and other structures) once the outer perimeter was breached—an example of "destruction in the name of construction" (Reddy 2012).

As another example, consider one of Tipu Sultan's armouries which lies about 700 m south-east of the Bangalore–Mysore railway line at Srirangapatna (Y in Fig. 6.8). Until recently, this structure was overgrown and buried under thick vegetation. We found cropmarks of this when we were studying Srirangapatna as part of a project funded by the Government of Karnataka in 2015–2016. We were

Fig. 6.7 A series Google Earth images of Chikkajala Fort in the course of widening National Highway 7 and a ground photograph of the ruins

convinced that the cropmark was one of Tipu Sultan's armouries, as the shape, size, and orientation resembled the one (X in Fig. 6.8) near the railway tracks and several others in the fort. We subsequently learned that there was a proposal to develop this property, which would have resulted in the loss of this structure. Fortunately, our report allowed the government to deny permission for construction and we later observed that the structure was conserved (Fig. 6.8). Of course, certain high-priority

Fig. 6.8 A series of Google Earth images of an armoury Y of Tipu Sultan at Srirangapatna along with ground photographs at different dates. For armoury X ref Fig. 6.9

development projects will be permitted despite the damage they will cause to built heritage, but a publicly accessible national-level database will at least enable informed debate to weigh the potential loss of cultural heritage against the advantages offered by the project, and ensure that such structures are not damaged or destroyed inadvertently.

The lack of such a database can delay decision-making for development projects, can significantly increase their costs, and can cause inconvenience to the public. As an example, the location of another of Tipu Sultan's armouries (Fig. 6.9) lay on the path of the planned doubling of the Bangalore–Mysore railway line at Srirangapatna. When the track-doubling project was proposed, a database that listed this historic structure would have alerted planners to the problem. At this early stage, the public could have been informed of the need to either demolish this historic structure or to consider alternatives with a range of associated costs. Unfortunately, the project was sanctioned and later stalled by the awareness of the structure. The government then proposed relocating the armoury at significant additional expense (Joshi 2015), when

Fig. 6.9 A series of Google Earth images of another armoury of Tipu Sultan at Srirangapatna located near the railway station. The ground photograph of the armoury was taken post-relocation

it was too late to consider less expensive or less disruptive alternatives. Figure 6.9 shows three Google Earth images. Figure 6.9a from February 2005 shows the partially buried armoury adjacent to the single railway line. Figure 6.9b is from February 2017, when the relocation was underway. The new location was prepared (by early March 2017)[16], and Fig. 6.9c from November 2017 shows the armoury in its new location together with the doubled railway line.

Authorities in charge of acquiring land for protecting built heritage and in enforcing protection against encroachment often face property-related disputes. In some instances, a geospatial database could assist authorities in resolving such disputes because satellite images can be used as evidence (Jadhav 1995). For instance, satellite images from previous years (or even previous decades in the case of Corona images) can help verify claims of historical land usage.

[16]https://www.youtube.com/watch?v=rminD6QSQdk; https://www.deccanherald.com/content/597983/tipus-armoury-shifting-goes-smooth.html; and https://timesofindia.indiatimes.com/city/mysuru/tipus-armoury-shifted-from-present-location/articleshow/57501816.cms. Accessed on 25 April 2020.

A geospatial database can also assist with site management. For instance, the development of infrastructure for visitors (e.g. information centres and parking areas) would need to avoid damaging features that lie above and below the ground, such as fort walls, mounds, moats, and palaeochannels. Further, the database can help determine subtler impacts to heritage sites from development activities in the surrounding area. As an example, Badami and Chitradurga forts are located amidst dramatically undulating rocky terrain intermingled with open rural landscape (Rajani et al. 2009; Nalini and Rajani 2012). The construction of tall structures such as mobile towers and ropeways can greatly detract from the aesthetic appeal of the site. In such instances, the database can be used to create 3D landscape models into which proposed developments can be embedded and visualized prior to authorizing them.

Once site protection boundaries are set, they are a double-edged sword: they protect the structures within while simultaneously heightening the threat to structures outside from redevelopment projects aimed at catering to the growing number of visitors. Hence, if the database is to assist in defining meaningful site protection boundaries, it must be populated with as exhaustive a list of potential locations of built heritage in the vicinity as possible. The onus for this task must rest with the academic research community, which has resources to conduct the necessary investigations (including geospatial analysis) and the capacity to evaluate the merits for the inclusion of specific locations based on a peer review of objective facts. Thus, the final and perhaps most critical purpose that such a database can serve is as an essential bridge connecting researchers with authorities who define and protect these boundaries. In the present situation, where no such connection exists, it is possible for the type of research presented in this book to be reduced to merely an academic exercise with no real-world consequences. We envisage a future in which well-researched contributions to such a database are encouraged and valued by all stakeholders.

References

Asher F (2015) Nalanda: situating the Great Monastery. The Marg Foundation

Asher F (2020) Sarnath: a critical history of the place where Buddhism began. The Getty Research Institute, Los Angeles

Cunningham A (1871) Archaeological Survey of India: four Reports made during the years 1862–63-64-65. Simla 1. (https://archive.org/details/fourreportsmade00cunngoog). Accessed 21 Apr 2020

Cunningham A (1892) Mahabodhi, or the Great Buddhist temple at Buddha-gaya. London, Plate 1. https://archive.org/details/cu31924008747788/page/n103. Accessed 24 Apr 2020

Gupta E, Das S, Suganya K, Kumar V, Rajani MB (2017) The need for a National archaeological database. Geospatial techniques in archaeology. Curr Sci Special Section 113(10):1961–1973 (Nov 25th)

Jadhav RN (1995) Encroachment in Sanjay Gandhi National Park. J Indian Soc Remote Sens 23:87–88

Joshi B (2015) US Company to move Tipu Sultan's 18th century armoury in single piece. Economic Times. May 29. http://timesofindia.indiatimes.com/india/Tipus-armoury-will-roll-across-to-new-location/articleshow/47579945.cms. Accessed 25 Apr 2020

Misra BN (1998) Nalanda, 3 vols. B. R. Publishing, Delhi

Nalini NS, Rajani MB (2012) Stone fortress of Chitledroog: visualizing old landscape of Chitradurga by integrating spatial information from multiple sources. Curr Sci 103(4):381–387 (25 Aug)

Rajangam K, Rajani MB (2017) Applications of geospatial technology in management of cultural heritage sites—potentials and challenges for the Indian region. Navalgund RR, Korisettar R (eds) Geospatial techniques in archaeology. Curr Sci Special Section 113(10):1948–1960 (25 Nov)

Rajani MB (2016) The expanse of archaeological remains at Nalanda: a study using Remote Sensing and GIS'. Archives of Asian Art 66(1):1–23

Rajani MB, Das S (2018) Archaeological remains at Nalanda: a spatial comparison of nineteenth century observations and the protected world heritage site. In: Sharma A (ed) Records, recoveries, remnants and inter-Asian interconnections: decoding cultural heritage. ISEAS–Yusof Ishak Institute, pp 239–256

Rajani MB, Patra SK, Verma M (2009) Space observation for generating 3D perspective views and its implication to the study of the archaeological site of Badami in India. J Cult Heritage 10(1):e20-e26. https://doi.org/10.1016/j.culher.2009.08.003 (Dec)

Reddy YM (2012) 3,000-year-old site in Bangalore is history, Bangalore, DNA (Daily News and Analysis), April 14th 2012. http://www.dnaindia.com/bangalore/report-3000-year-old-site-in-bangalore-is-history-1675707. Accessed 25 Apr 2020

Shashidhar K (2019) https://www.forbesindia.com/article/special/forbes-india-investigation-indias-most-gerrymandered-constituencies/53011/1. Accessed 23 Apr 2020

Stewart ML (1989) Nalanda Mahavihara: a study of an Indian Pala period Buddhist site and british historical archaeology, 1861–1938, BAR international series 529. B.A.R, Oxford

Chapter 7
Opportunities Beyond Landscapes

Check for updates

Preamble

This book has focused on geospatial analysis of archaeological landscapes, but such analyses can have implications for researchers in other disciplines as well. In this final chapter, we discuss three examples of such research linkages. In Sects. 7.1–7.3, we present three examples from our own research experience. Finally, in Sect. 7.4, we present an example where we see immense potential for research that can positively impact the protection of built heritage.

7.1 History of Astronomy

Astronomy has been closely intertwined with religion through much of history, and the construction of sacred structures was sometimes based on astronomical principles. For instance, the Egyptians may have aligned the Great Pyramid at Giza to cardinal directions using the circumpolar stars Kochab (Beta Ursae Minoris) and Mizar (Zeta Ursae Majoris) (Spence 2000). Hindu temples in India were also aligned to the cardinal directions, using other methods including Indian Circle (first described before 200 BCE in a treatise on mathematics and geometry called *Katyayana Sulbasutra*) (Sen and Bag 1983). The mathematician and astronomer Brahmagupta (born in 598 CE) noted that the Indian Circle method produced at most 0.5° of error in alignment) and Sripati, another mathematician and astronomer, suggested a correction for this error in 1039 CE or 1056 CE (Mollerup 2012).

Some contemporary Buddhist sacred structures such as the main stupas at Ratnagiri and Udayagiri (located 7 km apart in Orissa) are also aligned to the cardinal directions, but when we conducted a geospatial analysis, we noted that all the temples at Nalanda as well as other Buddhist temples within the same cultural milieu—the

M. B. Rajani, *Patterns in Past Settlements: Geospatial Analysis of Imprints of Cultural Heritage on Landscapes*, Springer Remote Sensing/Photogrammetry,
https://doi.org/10.1007/978-981-15-7466-5_7

Mahabodhi at Bodhgaya and central structures in Vikramasila, both in Bihar, India, Somapura in Bangladesh and Samye in Tibet—were oriented more than 4° south of true east. While some sacred structures are aligned to the rising sun, this hypothesis did not explain our observations. On consulting literature from the history of astronomy, we found that several ancient structures are believed to have been oriented towards the rising or setting of certain bright stars including Sirius (the Horus temple on the summit of Djebel Thoth in Western Thebes and the Isis temple at Dendera in Egypt) (Shaltout and Belmonte 2005; Belmonte et al. 2010), Antares (the temple of the Hurlers in Liskeard in England and the Older Erechtheum in Athen, Greece) (Brown 2000; Lockyer 1894[1]), and Spica (the temple of Min in Egypt, and several temples in Greece) (Olcott 1911). Relying on this scholarship, we hypothesized that the nine Buddhist temples under consideration were oriented towards the rise of a star. To identify this star, it was necessary to measure the orientations of the temples accurately.

It is challenging to measure these orientations at ground level for two reasons. First, many of these structures have been restored, so the orientation of a short wall segment may differ from the orientation of the overall structure. Unfortunately, it is difficult to find intact segments of straight walls that are several metres in length. Second, the Mahabodhi and Samye are functioning temples with many ancillary structures built around them, which makes it difficult to make accurate measurements. We therefore relied on high-resolution satellite imagery, which has been effectively used for measuring the orientation of ancient structures in Egypt and China (Shaltout 2014; Klokocník and Kostelecký 2010; Klokocník et al. 2011).

On Google Earth images, we identified and measured the orientations of straight lines at least 30 m long at each temple that align with surviving remnants of an east-west wall or the plinth. Since satellite images can be georeferenced, an added advantage is that geographical north can be established more accurately on these images than at ground level. To minimize measurement error at each site, we examined all pertinent Google Earth images and selected the ones closest to the nadir or direct vertical view (images taken at oblique angles distort the geometry). Our selections were determined visually, based on the parallax created by tall buildings present in the near vicinity of each site. To estimate the orientation of each structure, we identified two near-parallel linear features (walls or plinths) that flank it symmetrically to its north and south and then measured their east-west orientations using AutoCAD.

With these careful measurements, we were able to identify two candidate stars as the target for orienting these nine temples: either Spica or Beta Librae (Rajani and Kumar 2019). Although the latter star is not as bright as Spica, it is part of the Vishaka Nakshatra (lunar mansion) associated with the Buddha's birth. Since there is no prior evidence from the history of astronomy that Indian structures may also have been oriented towards the rise of stars, our interdisciplinary research has advanced our understanding in both domains. We believe the basis of orientation we have

[1]The methods used by Lockyer for data collection and analysis have been questioned by more recent authors (Papathanassiou 1994; Boutsikas and Ruggles 2011).

proposed has far broader applicability, not only to temples in India but even in other parts of the world where orientation was determined in relation to a specific star. We cannot, however, say whether structures distant from the Indian subcontinent used a similar mode of orientation (Rajani and Kumar 2019). This would be an interesting avenue for interdisciplinary research in future.

7.2 Military History

In the military context, maps can be created either prior to a battle (for planning purposes), or afterwards to analyse successful and unsuccessful decisions made during the battle. This type of analysis can be supplemented by textual records such as accounts by designated military observers. While the battles themselves need not leave visible scars, these spatial records can be a valuable source of information to answer questions of interest to military historians, particularly because maps made for military purposes often use the most accurate map-making techniques of their time.

We now describe how geospatial analysis was used to understand the range of eighteenth-century Mysorean rockets. These rockets were effectively used by the Mysorean army during the Anglo-Mysore wars (1780–82, 1790–92 and 1798–99), which pitted Hyder Ali and his son Tipu Sultan against the British East India Company. Although the British were the eventual victors, they were so impressed by these rockets that Colonel William Congreve led a vigorous technology programme to analyse them in Britain. In 1801–02, Congreve's tests confirmed that these Mysore rockets had more than double the range of the rockets available to the British. The estimates for the range vary according to the type of the rocket from 1000 yards (915 mts) to 1.5 miles (2.4 km). This outstanding performance was chiefly attributable to the iron employed for the casing, whereas European rockets were encased in wood or pasteboard. The high quality of the iron made this the first successful military deployment of such casings, which permitted increased bursting pressures and hence higher propellant packing density and therefore their impressive range (Narasimha 1999). These techniques were used to advance European rocketry with the development of the Congreve rocket in 1805 (Narasimha 1985).

Modern-day military historians seeking to validate the range achievable by Mysorean rockets could study their shape and dimensions: two samples of Tipu's rockets survive in the Royal Artillery Museum in Woolwich, UK, and a large number of casings were discovered recently in Nagara, Karnataka (Shejeshwara and Olikara 2018). However, there are numerous unknowns relating to how these rockets were prepared and fired that makes it challenging to confirm their 2.4 km range.

Fortunately, there is a rich historical spatial record of key battles that can be used to address this question. The events that took place at Srirangapatna on the night of 4–5 April 1799 have been particularly well documented by British soldiers who were part of this event (or by their biographers). This level of detail may partly be because the British were genuinely surprised by the capability of the Mysorean rockets, and partly because several notable military figures participated in this battle, including George Harris, David Baird, Lachlan Macquarie (Lushington 1811; Fortesque 1906; Hook 1832; Anonymous 1852; Macquarie archives[2]), and a young Arthur Wellesley. The latter, who later became known as the Iron Man of Britain and the Duke of Wellington, was reportedly "shell shocked" and disoriented by the sudden attack and explosions caused by Tipu's rockets at the Sultanpettah *Tope* (groves) where his army was positioned (Narasimha 1985).

By georeferencing the map *Seringapatam 1799*,[3] we geographically located the Sultanpettah Tope, the position of the Mysore army (from where the rockets were presumably launched), and the postings of British armies headed by Wellesley, Harris, Stuart, Hart, Shawes, and MacDonald (Fig. 7.1a). This map shows several built features that are identifiable on Google Earth images (and hence serve as good GCPs), including the fort, the gates, an aqueduct, a hill-temple, and several bridges. Features from the georeferenced map, including the Sultanpettah Tope, were then transferred to Google Earth imagery to find their current context (Fig. 7.1b).

With this georeferenced map, we were able to measure the distance between the nearest positions of the Mysore army to the easternmost edge of the Sultanpettah Tope. We were thus independently able to confirm that the range of these rockets was at least 2.48 km (Fig. 7.1b).[4]

7.3 Riverine and Coastal Geomorphology

The shapes of rivers and coastlines can vary over short durations (e.g. due to seasonal changes) and over longer durations (e.g. due to climate change). By identifying palaeochannels and strandlines detected in satellite images, we can sometimes determine the location and shape of a river or a coastline at some time in its past. In order to understand the geomorphology, we may want to determine how long ago this was. Typically, we try to find suitable sedimentary material for dating at the past location, but it can be challenging (or impossible) to select the right material. If the changes have occurred over archaeological timescales and there are built structures whose association with these past locations of rivers and coasts can be established using geospatial analysis, then it is sometimes easier to date these structures (e.g. using historical records or by analysing bricks or wood).

[2]Macquarie University Library in Australia https://www.mq.edu.au/macquarie-archive/seringapatam/intro.html Accessed 13 May 2020.

[3]Maps and plans illustrating Fortesque's History of the British Army (1906) 4:27.

[4]I am grateful to Prof Roddam Narasimha and Prof H.S. Mukunda for the many discussions about Mysorean rockets.

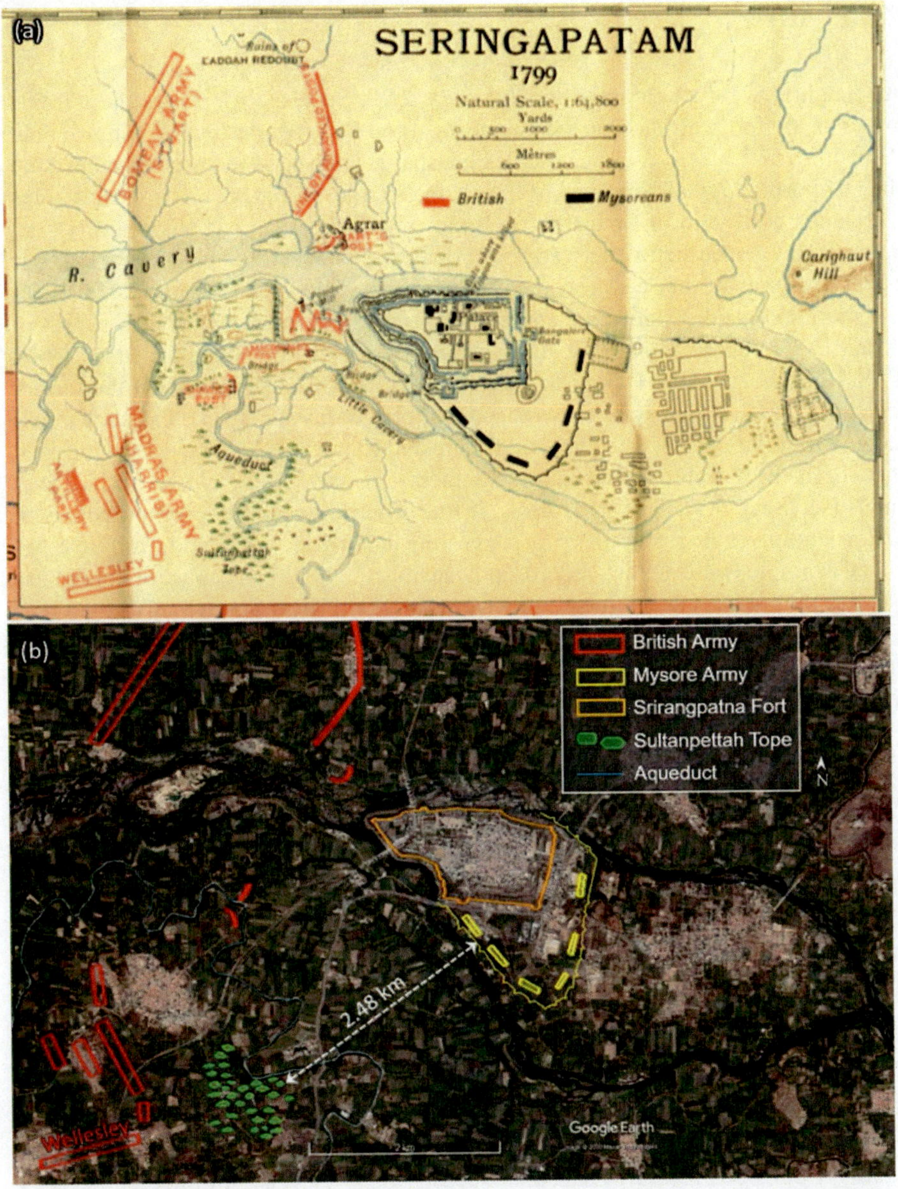

Fig. 7.1 **a** Map showing the postings of British and Mysorean armies in the vicinity of Srirangapatna in 1799 (reproduced from Fortesque 1906). **b** Features from the map in (**a**) georeferenced and overlaid on Google Earth imagery, showing the minimum distance between the position of the Mysore army and the Sultanpettah Tope

7.3.1 Rivers

A straightforward example is Sravasti (Fig. 2.25, Sect. 2.3.6), where the tell-tale curve of the fort wall records the past flow of the River Rapti (which is now about 2 km north of the fort). A more complex geospatial analysis is required to understand the geomorphology of Kosi, one of the Ganga's largest tributaries. It presently joins the Ganga north-east of Bhagalpur in Bihar (just west of the site Vikramasila), but an analysis of satellite imagery and historical maps of the region suggests that this confluence was further east less than two centuries ago (Chakraborty et al. 2010). Geomorphological insights can sometimes be made even for prehistoric sites. For instance, a comparison of the spatial distributions of mature Harappan sites (2200–1700 BCE) versus later Harappan sites (1700–1500 BCE) suggests that the former lay along a major river system (perhaps the fabled Sarasvati River). Palaeochannels suggest that this system dried up, and Rajani and Rajawat (2011) have hypothesized that the lack of water forced later settlements to move west, closer to the River Indus (see Sect. 4.1). This hypothesis, if true, contributes to the geomorphological understanding of the river system by suggesting that it dried up around 1700 BCE. As a final example, while textual records indicate that the Mauryan capital of Pataliputra (see Sect. 4.3.4) was at confluence of River Ganga and River Sone (McCrindle 1877; Cunningham 1871), its precise location is not known. Evidence for its location will enhance both our archaeological understanding as well as our understanding of the morphology of the River Ganga and its tributaries.

7.3.2 Coasts

One reason for dynamic changes to coastlines is sea level variations. We have data on sea levels recorded by tidal gauges from about 1700 CE, but data prior to this is at geological timescales and is typically estimated by extrapolating or interpolating from available data points (Church et al. 2001, 2013; Grinsted et al. 2010).

Auriemma and Solinas (2009) have suggested that careful analysis of archaeological data can offer important clues for understanding the evolution of coastlines in the Mediaeval Period, and Sundaresh et al. (2014) have demonstrated how evidence from coastal archaeological sites can be used to trace changes in coastlines with some accuracy. In addition, we believe that mediaeval and colonial period maritime maps are a relatively untapped source of data about coasts and sea levels. Prior to the industrial era, travel by water was often easier than overland. As a result, marine channels, navigable rivers, and sea crossings were important trade routes for ancient civilizations. This commercial endeavour led to the creation of Portolan charts (the Italian adjective *portolano* means "related to ports or harbours"). These charts were first made in Europe from the thirteenth century CE onwards, and they identified relatively permanent coastal features near ports and harbours (such as forts, temples, and hills) that were visible to sailors (Rajani and Kasturirangan 2013). Such charts, together with textual descriptions, have been published as collections and have contributed to our understanding of highly dynamic coastal features such as the growth of spit bars (Blake 2004; Boer and Carr 1969; Oldham 1925). Boer and Carr (1969) have suggested the use of such charts as a source of valuable information when other kind of evidence is sparse.

In Sect. 5.5, we have described how a seventeenth-century Dutch Portolan chart helped to explain why Mahabalipuram had the name Seven Pagodas. Our analysis technique simulated changes in the sea level to match the coastline depicted in the Portolan chart (Fig. 2.28). This match was further validated by agreement with a palaeostrandline identified in multispectral satellite imagery. By dating this strandline, our research has made a small contribution to understanding the morphology of this coast. We are presently investigating the potential of this technique to identify relative sea levels in the colonial period (sixteenth to twentieth century CE) in a systematic exploration of the entire coastline of India. The results of such studies should provide further data on the shape of past coastlines to assist geomorphologists, oceanographers, and for climatologists in refining existing coastline evolution models. In addition, this line of research can also help in identifying lost or submerged coastal archaeological sites, and in identifying sites that are vulnerable to coastal dynamics.

7.4 The Economics of Identifying and Protecting Built Heritage

Traditionally, identifying the remains of the past and preserving them have both been costly tasks. The first task has conventionally relied on laborious and time-consuming exploration and excavation, but the costs associated with geospatial analysis using remote sensing and GIS technologies have decreased rapidly with the increasing availability of online geoportals and image resources, together with reducing costs of remote sensing data.[5] Further, such analysis is typically far less destructive (Sect. 5.4.5—Talakadu). As noted in Sect. 6.2, these technologies can also be used to define more rational site protection boundaries and can help in enforcing them, potentially reducing the cost of the second task. These trends are expected to continue as the underlying technologies advance.

We close by reflecting on the implications of these trends on the economics of identifying and protecting built heritage in a country like India. Today, the Archaeological Survey of India has identified and protected over 3600 sites, and each state government protects a few hundred more. Some funds can be efficiently utilized to officially recognize thousands of unknown and undocumented remains. For each of these sites, modest additional resources are needed to define site protection boundaries, to raise public awareness about the importance of these sites, and to support

[5] https://www.nrsc.gov.in/sites/default/files/inline-images/Satellite_Data_Price_List.pdf Accessed 29 Apr 2020.

community-led initiatives to protect these remains (including by creating a public database of these sites and their boundaries, as noted in Sect. 6.3). More expensive forms of enforcing protection boundaries can be reserved for a limited number of sites.

Thus, geospatial analysis is likely to spur new research in understanding the economics of preserving our past. As we noted in a recent talk,[6] it is undeniable that we risk losing some of our cultural heritage due to lack of resources, but losing it due to ignorance—particularly, as the cost of dispelling that ignorance falls—is unacceptable.

References

Anonymous (1852) The military history of Duke of Wellington in India, London 1852. https://books.google.co.in/books?id=PFUBAAAAQAAJ&pg=PP5#v=onepage&q&f=false Accessed 28 May 2020

Auriemma R, Solinas E (2009) Archaeological remains as sea level change markers: a review. Quatern Int 206:134–146

Belmonte JA, Fekri M, Abdel-Hadi YA, Shaltout M, González-García AC (2010) On the orientation of ancient Egyptian temples: (5) testing the theory in middle Egypt And Sudan. J Hist Astron 41(1):65 (Feb)

Blake J (2004) The sea chart: the illustrated history of nautical maps and navigational charts. Conway Maritime Press, London

Boer GD, Carr AP (1969) Early maps as historical evidence for coastal change. Geogr J 135(1):17–27 (Mar)

Boutsikas E, Ruggles C (2011) Temples, stars, and ritual landscapes: the potential for archaeoastronomy in ancient Greece. Am J Archaeol 115(1):55–68 (Jan)

Brown PL (2000) Megaliths, myths and men: an introduction to astro-archaeology. Dover Publications, Mineloa, pp 75–76 (17 March)

Chakraborty T, Kar R, Ghosh P, Basu S (2010) Kosi megafan: historical records, geomorphology and the recent avulsion of the Kosi River. Quat Int 227(2010):143–160 (Elsevier)

Church JA, Clark PU, Cazenave A, Gregory JM, Jevrejeva S, Levermann A, Merrifield MA, Milne GA, Nerem RS, Nunn PD, Payne AJ, Pfeffer WT, Stammer D, Unnikrishnan AS (2013) Sea level change. climate change: the physical science basis. In: Stocker TF, Qin D, Plattner GK, Tignor M, Allen SK, Boschung J, Nauels A, Xia Y, Bex V, Midgley PM (eds) Contribution of working group I to the fifth assessment report of the intergovernmental panel on climate change. Cambridge University Press, Cambridge

Church JA, Gregory JM, Huybrechts P, Kuhn M, Lambeck K, Nhuan MT, Qin D, Woodworth PL, Anisimov OA, Bryan FO, Cazenave A, Dixon KW, Fitzharris BB, Flato GM, Ganopolski A, Gornitz V, Lowe JA, Noda A, Oberhuber JM, Farrell SP, Ohmura A, Oppenheimer M, Peltier WR, Raper SCB, Ritz C, Russell GL, Schlosser E, Shum CK, Stocker TF, Stouffer RJ, van de Wal RSW, Voss R, Wiebe EC, Wild M, Wingham DJ, Zwally HJ (2001) In: Houghton JT, Ding Y, Griggs DJ, Noguer M, van der Linden PJ, Dai X, Maskell K, Johnson CA (eds) Climate change 2001: the scientific basis. The Press Syndicate of the University of Cambridge. http://cedadocs.ceda.ac.uk/981/13/Chapter_12.pdf.. Accessed 29 Apr 2020

Cunningham A (1871) Reports of tours in the gangetic provinces from Badaon to Bihar in 1875–76 and 1877–78, vol 11. Printed at the Government Central Press, Simla

Fortesque JW (1906) A history of the British Army, London

[6]https://www.ted.com/talks/m_b_rajani_clues_in_the_landscape Accessed 30 Aug 2020.

Grinsted A, Moore JC, Jevrejeva S (2010) Reconstructing sea level from paleo and projected temperatures 200 to 2100 AD. Clim Dyn 34:461–472. https://doi.org/10.1007/s00382-008-0507-2

Hook TE (1832) The life of general the right honourable Sir David Baird, London. https://books.google.co.in/books?id=Tijrm3XMV2AC&pg=PP1#v=onepage&q&f=false. Accessed 28 May 2020

Klokocník J, Kostelecký J (2010) Google earth for the study of ancient civilizations. In: Second international conference on advanced geographic information systems, applications, and services, pp 56–61

Klokocník J, Kostelecký J, Pavelka K (2011) Google earth: inspiration and instrument for the study of ancient civilizations. Geoinform CTU FCE 6:193–211

Lockyer JN (1894) The dawn of astronomy; a study of the temple-worship and mythology of the ancient Egyptians, London, p 419. https://archive.org/details/dawnastronomyas00lockgoog/page/n438. Accessed 24 May 2019

Lushington SR (1811) The life and service of General Harris: during his campaigns in America. The West Indies and India, London

McCrindle JW (1877) Ancient India as described by Megasthenes and Arrian. Indian Antiquary. https://archive.org/details/in.gov.ignca.34982. Accessed 28 Apr 2020

Mollerup A (2012) Orientations of Khmer temples & the Indian circle. Ancient Khmer sites in Eastern Thailand. White Lotus Publishing, Bangkok, pp 147–160

Narasimha R (1985) Rockets in Mysore and Britain, 1750–1850 A.D. National Aeronautical Laboratory, Bangalore. https://www.researchgate.net/publication/37179995_Rockets_in_Mysore_and_Britain_1750-1850_AD. Accessed 28 May 2020

Narasimha R (1999) Rocketing from the Galaxy Bazaar. Nature 400. https://www.nature.com/articles/22022.pdf?origin=ppub. Accessed 29 May 2020 (8 July)

Olcott WT (1911) Star lore of all ages; a collection of myths, legends, and facts concerning the constellations of the Northern Hemisphere. G.P. Putnam's sons, New York, p 387. https://archive.org/details/starloreofallage00olco/page/n10/mode/2up. Accessed 5 May 2020

Oldham RD (1925) The Portolan maps of the Rhône Delta: a contribution to the history of the sea charts of the middle ages. Geogr J 65(5):403–424 (May)

Papathanassiou (1994) Archaeoastronomy in Greece: data, problems and perspectives, pp 433–443

Rajani MB, Rajawat AS (2011) Potential of satellite based sensors for studying distribution of archaeological sites along palaeo channels: Harappan sites a case study. J Archaeol Sci 38(9):2010–2016 (Elsevier, Sept). https://doi.org/10.1016/j.jas.2010.08.008

Rajani MB, Kasturirangan K (2013) Sea level changes and its impact on coastal archaeological monuments: seven pagodas of Mahabalipuram, a case study. J Indian Soc Remote Sens 41:461–468. https://doi.org/10.1007/s12524-012-0210-y

Rajani MB, Kumar V (2019) Nalanda: a tale in the twist. J Soc Architectural Historians 78(4):392–408. https://doi.org/10.1525/jsah.2019.78.4.392 (Dec)

Shejeshwara R, Olikara NG (2018) Rockets from Mysore under Haidar Ali and Tipu Sultan: preliminary studies of 'Tipu rockets' from the nagara find. J Arms Armour Soc 22(6):332–355

Sen SN, Bag AK (trans) (1983) The Sulbasutra of Baudhayana, Apastamba, Katyayana and Manava. Indian National Science Academy, New Delhi

Shaltout M (2014) Studying the orientations of luxor ancient Egyptian temples using QuickBird images. J Earth Sci Eng 4:194–210

Shaltout M, Belmonte JA (2005) On the orientation of ancient Egyptian temples I: upper Egypt and lower Nubia. J Hist Astron 36(3):273–298 (July)

Spence K (2000) Ancient Egyptian chronology and the astronomical orientation of pyramids. Nature 408:320–324 (16 Nov)

Sundaresh, Mani MR, Seelam JK, Gaur AS (2014) Shoreline changes along Tamil Nadu coast: a study based on archaeological and coastal dynamics perspective. Indian J Mar Sci 43(7) (July)

Index of sites

M. B. Rajani, *Patterns in Past Settlements: Geospatial Analysis of Imprints of Cultural Heritage on Landscapes*, Springer Remote Sensing/Photogrammetry,
https://doi.org/10.1007/978-981-15-7466-5

Index of terms

M. B. Rajani, *Patterns in Past Settlements: Geospatial Analysis of Imprints of Cultural Heritage on Landscapes*, Springer Remote Sensing/Photogrammetry,
https://doi.org/10.1007/978-981-15-7466-5